Barcode in Back

Teamology: The Construction and Organization of Effective Teams

Douglass J. Wilde

Teamology:
The Construction
and Organization
of Effective Teams

 Springer

Douglass J. Wilde, Professor Emeritus
Design Division
Department of Mechanical Engineering
Stanford University
Building 530, 440 Escondido Mall
Stanford, CA94305-3030
USA

ISBN 978-1-84800-386-6   e-ISBN 978-1-84800-387-3

DOI 10.1007/978-1-84800-387-3

British Library Cataloguing in Publication Data
Wilde, Douglass J.
  Teamology: the construction and organization of effective teams
  1. Jung, C. G. (Carl Gustav), 1875–1961
  2. Group problem solving
  3. Cognitive psychology
  4. Typology (Psychology)
  5. Myers-Briggs Type Indicator
  6. Group work in education – Case studies
  I. Title
  302.3
ISBN-13: 9781848003866

Library of Congress Control Number: 2008935392

Cover design: eStudio Calamar S.L., Girona, Spain

Printed on acid-free paper

9 8 7 6 5 4 3 2 1

springer.com

To Ariana Wilde,
my bright one and only grandchild.

Preface

This book is the product of sixteen years of studying student teams in engineering design project courses, mainly at Stanford University. The book shows how psychiatrist C. G. Jung's cognition theory, a cornerstone of modern personality typology, may be used to form and organize effective problem-solving teams. It does this through a novel quantitative transformation of numbers from the Myers–Briggs Type Indicator (MBTI) psychological instrument directly on to Jung's eight cognitive modes. The quantitative mode scores resulting make obvious what is needed to make a good team.

Using these methods will make an entire team project class perform as well as what would be, without them, the top quartile. This extravagant claim is based partly on experience with the ability of Stanford's graduate teams to win national design prizes. It also comes from direct observation of project courses small and large at Stanford and at other universities as far away as Shanghai.

The basic idea is to have every team possess among its members the full range of problem-solving approaches available to the human race. People who individually have only a few problem-solving strategies can pool these on a good team to make it overcome any obstacle it encounters. Since some of these strategies are introverted and thus not readily apparent to outside observers, a questionnaire is needed to discover what modes each person has to contribute. The best questionnaire for this is the Myers–Briggs Type Indicator (MBTI); many classes use the smaller 20-item version in the book's second chapter. The numbers from either of these questionnaires are transformed easily into the cognitive mode scores employed to construct and organize the teams.

An advantage of using such an objective questionnaire is its avoidance of the cliques and conformity that produce inferior teams. The MBTI also generates in team members the self-awareness and tolerance of different points of view needed on a cognitively diverse team. This diversity has brought about an expansion of conventional personality type theory that is sure to intrigue adherents of the MBTI.

Three audiences are targeted. First are educators in charge of engineering project courses, particularly the "capstones" of the senior year and first year introductory courses intended to motivate potential majors. This does not exclude non-academics such as project managers and human resource professionals. Extensions to other types of team – corporate, medical, literary or legal – would seem possible even when not discussed explicitly.

The second audience includes students and working professionals on project teams at all levels of Engineering, Architecture and Business. This does not exclude the rare course focusing on personal development through teamwork, a relatively new educational approach suggested by the author's experience with such an undergraduate seminar.

Third are the MBTI users and counselors interested in personal self-awareness and the development of interpersonal ability through the book's quantitative transformation of MBTI scores. This transformation into the Jungian domain yields "cognition patterns" – descriptions more detailed than those in the celebrated Myers–Briggs Type Table and better suited for the analysis of teams.

The book begins by recounting the remarkable improvement experienced by Stanford teams resulting from proper management of the various diversities a good team should have, especially psychological diversity as measured by the MBTI. It also summarizes Jung's cognition theory as an introduction to Chap. 2.

The second chapter, probably the one most interesting to MBTI professionals and Jungian analysts, discusses the measurement of Jung's cognition by the MBTI and how to transform the MBTI clarity scores rigorously into cognitive mode scores.

Chapter 3, especially of interest to faculty and corporate management, shows how the mode scores can be used to form uniformly excellent teams from a pool of students or working professionals. Although not at all deep mathematically, this chapter is no doubt the one requiring the most concentration by the "Team-meister" responsible for team formation. For top performance, computer assistance may be needed for large personnel pools.

Using mode scores to organize an existing team, no matter how formed originally, is developed in the fourth chapter. The organization meeting in which this is done has been found important for speeding up the team building process. This chapter will particularly interest people already on working teams.

Chapter 5 interprets modal scores in terms of individual personality. The personality descriptions generated will particularly interest MBTI adherents, for they furnish valuable expanded second opinions relative to the famous sixteen-element Myers Type Table.

Although much of the book concerns the normal psychology of Jung's personality theory, the author is a systems analyst rather than a psychologist. The relation between psychology and mathematics here is analogous to the fruitful one between physics and engineering. Well-known elements of psychology, as in physics, are organized in powerful new ways, providing at the same time practical applications and novel interpretations of the underlying science. The psychological elements of the theory, being widely known and broadly accepted, are reviewed but not challenged. It is the system in which the elements are imbedded, analyzed by simple but rigorous mathematics, that is strengthened to yield a new, dramatically expanded typology. This work may thus begin a productive new interdisciplinary activity – "teamology" – accessible eventually to system analysts and psychologists alike.

A glossary follows the last chapter. References in the text give author and publication date, with citation details collected at the very end of the book.

Many thanks to Stanford's Center for Design Research, especially to Professors Larry Leifer and Mark Cutkosky, as well as to Stanford's Office of Undergraduate Education, for encouraging and supporting this rather unconventional research. The very careful review of the previous draft's terminology by Dr. Judith Breimer, Research Director of the Center for the Application of Psychological Type (CAPT) in Gainesville FL motivated the blending of the new quantitative Jungian modal theory with the conventional Myers qualitative theory based on the MBTI letters alone.

Mechanical Engineering (Design) and Chemical Engineering
Stanford University
Stanford, CA, USA 94305

Doug (Douglass J.) Wilde
Professor Emeritus

Contents

Abbreviations and Notation

ag.	affinity group
ag.s	affinity groups
c-	pertaining to information **C**ollection
d-	pertaining to **D**ecision-making
E	extraverted, also extraversion clarity index
EC	extraverted information **C**ollection (attitude), also score
ED	extraverted **D**ecision-making (attitude), also score
EF	extraverted feeling (cognitive mode), also score
EF	extraverted feeling affinity group membership, regular
EF	extraverted feeling affinity group membership, marginal
E f	Diplomat (team role)
eF	Conciliator (team role)
EI	Questionnaire score for Energy Direction
EN	extraverted intuition (cognitive mode), also score
EN	extraverted iNtuition affinity group membership, regular
EN	extraverted iNtuition affinity group membership, marginal
En	Entrepreneur (team role)
eN	Innovator (team role)
ES	extraverted sensing (cognitive mode), also score
ES	extraverted sensing affinity group membership, regular
ES	extraverted sensing affinity group membership, marginal
Es	Tester (team role)
eS	Prototyper (team role)
ET	extraverted thinking (cognitive mode), also score
ET	extraverted thinking affinity group membership, regular
ET	extraverted thinking affinity group membership, marginal
Et	Coordinator (team role)
eT	Methodologist (team role)

F	feeling, also MBTI feeling "judgment" clarity index
I	introverted, also introversion clarity index
IC	introverted information **C**ollection (attitude), also score
ID	introverted **D**ecision-making (attitude), also score
IF	introverted feeling (cognitive mode), also score
IF	introverted feeling affinity group membership, regular
_IF_	introverted feeling affinity group membership, marginal
If	Critiquer (team role)
iF	Needfinder (team role)
IN	introverted iNtuition (cognitive mode), also score
IN	introverted iNtuition affinity group membership, regular
_IN_	introverted iNtuition affinity group membership, marginal
In	Strategist (team role)
iN	Visionary (team role)
IS	introverted sensing (cognitive mode), also score
IS	introverted sensing affinity group membership, regular
_IS_	introverted sensing affinity group membership, marginal
is	Inspector (team role)
iS	Investigator (team role)
IT	introverted thinking (cognitive mode), also score
IT	introverted thinking affinity group membership, regular
_IT_	introverted thinking affinity group membership, marginal
it	Reviewer (team role)
iT	Specialist (team role)
JP	Questionnaire score for Orientation
J	structured, also MBTI "judgment" clarity index
MBTI	Myers–Briggs Type Indicator
N	iNtuition, also MBTI iNtuitive "perception" clarity index
P	flexible, also MBTI "perception" clarity index
S	sensing, also MBTI sensing "perception" clarity index
SN	Questionnaire score for Information Collection
T	thinking, also MBTI thinking "judgment" clarity index
TF	Questionnaire score for Decision-making

Chapter 1
Diversified Teams

Adding a few people who know less, but have diverse skills,
actually improves the group's performance.
– James Surowiecki, *The Wisdom of Crowds*, 2004

1.1 Introduction

Described here is a way to use Jungian cognition theory to construct and
organize problem-solving teams. This opening chapter records the method's
documented successes with engineering student design teams, advises the
reader how to use the book, outlines the underlying approach, and discusses
why and how the method improves team performance.

This method for composing and analyzing "cognitively diverse" engineering student design teams has been developed since 1991 in Stanford University's Mechanical Engineering Design Division. In its flagship graduate
course, student teams design, build, demonstrate and present projects motivated by industrial, medical and social problems (Wilde 1997). Application
of this method tripled the fraction of Stanford teams awarded Lincoln Foundation graduate design prizes. Further study showed that team quality can
be advanced even further – up to an equivalent of bringing ALL teams to
championship level – by having the teams organize themselves according to
the information generated in forming their teams.

1.1.1 Tripling the Prize Frequency

Figure 1.1 graphically illustrates the improvement during the first decade.
The leftmost bar shows average performance in the years before any version

Douglass J. Wilde, *Teamology: The Construction and Organization
of Effective Teams* © Springer 2009

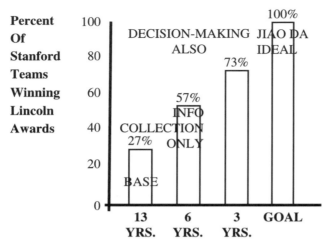

Fig. 1.1 Improvement history of Stanford teams

of the method was employed. The first six-year trial exhibited the two-fold improvement shown in the next bar. The third bar shows that the performance using a simpler version of the methods of Chap. 3 tripled during the next three-year trial, after which no further award records were kept. The rightmost bar represents the highest effectiveness theoretically possible. Chapter 3 suggests ways to approach such ideal performance through focused computation.

1.1.2 Achieving the Ideal

Improving three things actually did bring the ideal situation. First, an error in computation was corrected. Second, a method for producing maximum variation was developed. Third a good procedure for getting each team organized was necessary. When these developments were tested on 40 freshman teams at Shanghai Jiao-Tong (Jiao Da) University in 2007, almost all teams performed at prize level.

1.1.3 Evaluation: The Lincoln Design Awards

A few words are in order about the Lincoln Awards because the remarkable increased success of Stanford teams in winning them is advanced here as anecdotal evidence of the effectiveness of the team formation method. Since the late 1940s the Lincoln Arc Welding Foundation of the Lincoln Electric Company has annually awarded twenty-four prizes – a dozen graduate and a dozen undergraduate – to designs produced by U. S. student teams, regardless of discipline. The judging is based on reports submitted by the teams. All university identification is deleted from the reports, which are read by an anonymous panel, renewed each year, of prominent designers and design professors selected by the foundation. The panel made twelve cash awards at each student level: one "Best of Program", one Gold, two Silver, three Bronze, and five Merit.

To gain an independent evaluation of its own program, the Design Division of Stanford's Mechanical Engineering Department has been submitting all reports from its graduate design course to the Lincoln Foundation since 1979. As the preceding figure shows, during the first years about a quarter of the Stanford teams won Lincoln awards, which the faculty found quite satisfying considering the national competition. The striking improvement over the decade following, when the methods of this book were applied, is recorded in Fig. 1.1. Though not scientific proof of the method's effectiveness, this is strong enough anecdotal evidence to merit documentation of the method here for the use of other design teams, as well as for further research.

1.1.4 Confirmation Studies

Several universities have conducted statistical studies of these team construction methods. At the University of California at San Diego, Mechanical Engineering professor Nathan Delson has subjected his sophomore design course for engineering and biology students to double-blind statistical research by Joan Connell of the Psychology faculty. In the autumn of 2004, half of Delson's thirty-six student teams, mostly quartets with a few trios, were formed according to the "cognitive diversification" methods of Chap. 3, the other students being assigned at random. Only Connell, neither Delson nor the students, knew how particular teams were formed. Team robotic projects, many involved in contests, were evaluated by Delson, elements of the team behaviors being studied through student responses to questionnaires. The two statistically significant findings were that, although the "cognitively diverse" teams felt less cohesive to their members, their projects were indeed judged more creative by Delson.

At the University of Florida, Industrial Engineering Professor R. Keith Stanfill tried out modal variety on nine industrial engineering teams using origami production to simulate industrial design and manufacture. The quality of product increased with the closeness to the modal variety model. That is, the two teams having all eight cognitive modes covered out-performed those with seven or fewer. Stanfill said that the quality was "almost in order of the number of modes" as predicted.

1.2 An Overview of this Book

1.2.1 Introduction

This section provides an overview of the book intended to guide readers – students, faculty, team members and supervisors – in deciding how to use it. Chapter 2 gives the questionnaire and shows how it generates a cognitive pattern. Chapter 3 describes how to form the kind of cognitively diverse teams that have proven so effective. Chapter 4 shows how to use the personality patterns computed in Chap. 2 to organize existing teams, no matter how constructed. Readers may find in Chap. 5 new understandings of their own personalities that reach beyond the famous MBTI type table, whether or not they happen to be on teams themselves at the moment. Finally, Chap. 6 proves everything rigorously and surveys related issues such as the applications to education and personal development.

As suggested in Surowiecki's epigraph to this chapter, variety is the key to improved team performance. There are four kinds of variety relevant, three of them easy to arrange. The fourth more subtle "cognitive" kind, the one that made the difference at Stanford, is the subject of this book. The four kinds of variety are: experiential, professional, sociological and finally, cognitive.

1.2.2 Experiential Variety

Experiential variety occurs when team members have different levels of professional experience. This is most noticeable in an academic course when inexperienced students mingle with older graduates having several years of practical experience. In this situation the faculty is wise to keep the experienced people from banding together to the obvious disadvantage of the inexperienced students. This can be done simply by identifying the more experienced and asking them not to team up together. If they comprise less than a quarter of the enrollment, the rule can be promulgated that no quartet should have more than one experienced member. For more than a quarter but less than a half the limit would be two, and so on for larger fractions. This approach is equally valid for non-academic project teams, but somebody in authority must publicize it and help the members enforce it.

1.2.3 Professional Variety

Professional variety involves having various professions or skills on a team. This is frequently made deliberate in an academic setting by opening a course to several majors, engineering and business or computer science for example. This can be handled in the same way as experiential variety, by identifying the various specialties and asking that all the major concentrations be represented on every team. The corporate version of this is not only obvious, it is the norm when teams must cross disciplinary lines, needing for example a computer person, a marketing whiz, a biomechanical engineer and a patent expert on board.

1.2.4 Sociological Variety

Sociological variety comes into play as non-traditional groups enter a profession, as women did a generation ago in engineering. There is a need to avoid restricting the education and integration of the newcomers by segregating them. This can also be a problem with overseas students having language problems. The solution is the same as for experiential and professional variety – request that each team have at least one member from the non-traditional group. When there are several subgroups, the rule would be merely to have the number of non-traditional members on every team in line with the overall proportion in the class. A short summary rule would be that no team be stereotypable as, say, the "Women's", "Asian" or "Foreign" team.

1.2.5 Cognitive Variety

The preceding three non-psychological varieties thus act as minor constraints on the range of choice of team members. These can be imposed on top of the more subtle demands of the "cognitive" variety to be described next. Cognitive variety is guided by the results of a questionnaire given in the next chapter for estimating personal preferences. Of the four varieties, the cognitive one is the most difficult to achieve for team construction. But it is well worth seeking because it is cognitive variety that seems to have generated the greatly improved performance of the Stanford teams.

The psychoanalyst C. G. Jung (1875–1961) (see Jung 1971) modeled the conscious part of the human personality in terms of eight "cognitive modes", each occurring at different consciousness levels in a given person. The eight modes are listed in Table 1.1. Each mode is described both by Jung's technical name for it, "extraverted sensing" abbreviated "*ES*" for example, and a capitalized keyword introduced here for clarification such as "EXPERI-MENT" for *ES*. The modes represent eight different approaches to solving a problem. An individual would not, of course, expect to have all eight modes immediately available. More likely, one mode would be a favorite, with some selectable in special circumstances and others hard to imagine using at all.

A team of several people, however, can bring multiple modes to bear on any problem simultaneously. This justifies a team's having members with different favorite modes, the condition known as "cognitive variety" that seems to explain the enhanced success of the diversified Stanford teams. Jung's first personality theory classified people only according to their favorite modes, although later his typology included modes that are less preferred.

The cells of Table 1.1 are arranged to aid understanding of the relations between the modes. The upper row shows the four "extraverted" modes (*ES*, *EN*, *ET*, *EF*) used in the world exterior to individuals, whereas the lower four "introverted" modes (*IS*, *IN*, *IT*, *IF*) are employed internally in ways

Table 1.1 Jung's cognitive modes

Extraverted Sensing	Extraverted iNtuition	–	Extraverted Thinking	Extraverted Feeling
ES	*EN*		*ET*	*EF*
EXPERI-MENT	**IDEATION**		**ORGANI-ZATION**	**COMMU-NITY**

Introverted Sensing	Introverted iNtuition	–	Introverted Thinking	Introverted Feeling
IS	*IN*		*IT*	*IF*
KNOW-LEDGE	**IMAGI-NATION**		**ANALYSIS**	**EVALUA-TION**

INFORMATION COLLECTION   DECISION-MAKING

rarely apparent to an outsider. Each pair of modes sharing a column employs exactly one of Jung's four mental "functions", namely "Sensing" (*ES*, *IS*), "iNtuition" (*EN. IN*), "Thinking" (*ET*, *IT*) and "Feeling" (*EF*, *IF*) from left to right. These functions describe how the extraverted or introverted energy is used. For example, *ES* extraverted sensing involves hands-on experimentation, whereas *EN* extraverted iNtuition is concerned with theoretical ideas about future results.

The four modes on the left (*ES*, *EN*, *IS*, *IN*) involve information collection; those on the right (*ET*, *EF*, *IT*, *IF*), decision-making. Within each foursome, modes diagonally opposite are said to be "complementary" because not only do the psychological functions differ, but also one mode is introverted and the other extraverted. Thus choosing one mode automatically relegates the other to a subordinate preference or even total rejection. For example, an Ideation *EN* person would favor novelty over using something already known the way an *IS* Knowledge person would. The other three complementary pairs are *ES* Experiment vs. *IN* Imagination, *ET* Organization vs. *IF* Evaluation, and *EF* Community vs. *IT* Analysis. This complementarity will show up in the measurement phase; consciousness scores for complementary modes will be numerically equal but have opposite signs. Hence scores of *EN* 6 and *IS* −6 are equivalent expressions of preference for Ideation *EN*. The arrows in Table 1.1 depict complementarity graphically.

1.2.6 Mode Measurement

Although Jung spoke of levels of consciousness varying from mode to mode for an individual, his theory was entirely qualitative. Chapter 2 will deal with the measurement of preference for one mode over its complement as a quantity increasing with, although not strictly proportional to, consciousness of the mode. The numbers generated, based on simple counts of responses to a questionnaire, are then used to guide team formation and organization.

Filling out and scoring the questionnaire has been computerized at Stanford to save class time, so for anyone with access to such a program, Chap. 2 is not really necessary except as background information on the origin and validity of the questionnaire. The scores are used in Chap. 3 to compose teams so that as much as possible each mode has someone preferring it. This approach assumes that such a person will perform well in putting the mode to work for the team. Secondarily, duplication of preference for any mode is minimized to reduce rivalry among team-mates. Such cognitive diversity is not easy to achieve in practice, largely because in most personnel pools some modes are more preferred than others.

Aside from its employment to form teams in Chap. 3, this quantitative cognitive mode information has other uses. The first is to help the team organize itself (Chap. 4), paying particular attention to roles for which there is no obvious leadership on board. This is used to describe the roles fitting various members of the team. The second use of the mode scores (Chap. 5) is the self-understanding that comes from a personality description based on the cognitive modes. This application should be of particular interest to coaches and counselors.

As the Stanford faculty gained experience with cognitively diverse teams, additional justifications for cognitive variety emerged that were independent of any empirical demonstration of improved prize-winning effectiveness. Three plausible reasons for these improvements are: development of individual competence, clear assignment of team responsibility, and increased opportunities for experience with very different people with whom one might otherwise not come in contact.

1.3 Concluding Summary

Using the preference questionnaire to construct teams has been shown to raise the fraction of design prize-winners at Stanford from one-quarter to three-quarters. Achieving 100% would seem to depend on employing the modal information as shown in Chaps. 3 and 4. This does not mean that all teams can be guaranteed to win prizes. It just means that all teams would produce results of prize-winning quality. Although each such team would have distinguishable differences, the overall quality of the finished project would be uniformly as high as possible – a design professor's dream! Now for the questionnaire and, later, how to use it for constructing and organizing top-notch teams.

Chapter 2
Questionnaire and Transformation

Know thyself.
– Inscription on the temple to Apollo at Delphi, ascribed to Solon

You can hardly expect to understand your working group or its members without understanding yourself first. This chapter intends to give you insight into your own preferences and potentials. The method described for achieving this is powerful enough to extend to the understanding of your colleagues, or even to the organization of your team as a whole (Chap. 4). It also lays a foundation for selecting a good new team from a large set of equally qualified candidates, a topic to be developed fully in Chap. 3.

2.1 The Cognitive Questionnaire

Stanford's computer input program, upon which the formation and organization of cognitively diverse teams is based, employs the twenty items of the questionnaire in Table 2.1. For more psychologically valid results the program randomizes the orders of both the questions and the responses.

2.1.1 Questionnaire

Here the questions and responses have been grouped to make discussion and scoring easier. Interpretations of the four categories will follow afterward.

If the input program is not at hand, you can use the questionnaire to estimate which modes you prefer – your "cognitive pattern" – and determine scores to be employed for placing you well on a team. Often this questionnaire has been used right in class, the filling out and scoring taking no more

Douglass J. Wilde, *Teamology: The Construction and Organization of Effective Teams* © Springer 2009

Table 2.1 Cognitive mode questionnaire

Circle zero, one or two alternatives for each of 20 questions.

Be careful with signs!

Energy Direction: Outward or Inward

EI1	You are more:	(e)	sociable	(i)	reserved
EI2	You are more:	(e)	expressive	(i)	contained
EI3	You prefer:	(e)	groups	(i)	individuals
EI4	You learn better by	(e)	listening	(i)	reading
EI5	You are more:	(e)	talkative	(i)	quiet

EI difference: $\Sigma e - \Sigma i = $ EI_____

Orientation: Structured or Flexible

JP1	You are more:	(j)	systematic	(p)	casual
JP2	You prefer activities:	(j)	planned	(p)	open-ended
JP3	You work better	(j)	with pressure	(p)	without pressure
JP4	You prefer:	(j)	routine	(p)	variety
JP5	You are more:	(j)	methodical	(p)	improvisational

JP difference: $\Sigma j - \Sigma p = $ JP_____

Information COLLECTION process: Facts or Possibilities

SN1	You prefer the:	(s)	concrete	(n)	abstract
SN2	You prefer:	(s)	fact-finding	(n)	speculating
SN3	You are more:	(s)	practical	(n)	conceptual
SN4	You are more:	(s)	hands-on	(n)	theoretical
SN5	You prefer the:	(s)	traditional	(n)	novel

SN difference: $\Sigma s - \Sigma n = $ SN_____

DECISION-making process: Objects or People

TF1	You prefer:	(t)	logic	(f)	empathy
TF2	You are more:	(t)	truthful	(f)	tactful
TF3	You see yourself as more:	(t)	questioning	(f)	accommodating
TF4	You are more:	(t)	skeptical	(f)	tolerant
TF5	You think judges should be:	(t)	impartial	(f)	merciful

TF difference: $\Sigma t - \Sigma f = $ TF_____

than half an hour. Even if you have used the input program, you may wish to fill out the questionnaire just to understand what is being measured and to double-check your program scores (Exercise 2.1).

The questionnaire asks you to choose between the two alternatives given for each item. For example, the first item is (Table 2.2):

Table 2.2 Typical questionnaire item

| *EI*1 | You are more: | (e) | sociable | (i) | reserved |

Most people will simply choose one or the other and circle the letter (e) or (i). But the choice also includes the option of selecting both alternatives if you really do use them with about the same frequency. You may also leave both blank if an item is unclear or if both choices seem unlikely, situations that might occur especially if English is not your native language.

2.1.2 Mind-Set

It would be mistaken to regard the questionnaire as a quiz or test of competence. It is rather an assessment of what you usually like to do, which of course may not be what is expected of you in most college courses, especially at examination time. It is better to record your natural preferences rather than what you might think your family or teachers expect of you. Being honest with yourself is most likely to get you a team assignment as close as possible to what you really like to do and are probably good at. Kidding yourself is asking for trouble.

2.1.3 Calculations

Each choice adds or subtracts a point or two to half or all of the eight mode scores in the squares of Table 2.3. But rather than compute item-by-item, just add up the responses within each of the top four sections and record

Table 2.3 Cognitive mode scores

Compute and record **non-negative** scores below.

EI - JP	EI - JP		EI + JP	EI +JP
+ 2SN =	− 2SN =	-	+ 2TF =	− 2TF =
ES____	EN ____		ET ____	EF____
EXPERI-MENT	IDEATION		ORGANI-ZATION	COMMU-NITY
- EN =	- ES =	-	- EF =	- ET =
IS ____	IN ____		IT ____	EF____
KNOW-LEDGE	IMAGI-NATION		ANALYSIS	EVALU-ATION

INFORMATION Collection   Decision-Making

the differences where indicated. The algebraic signs are crucial, so follow the difference formulas carefully. For example, 2 (e) responses and 3 (i) responses would give an EI difference of $2 - 3 = -1$.

Each mode has a three-term formula giving the score for the mode. For example, the upper left "Experiment" mode has the equation $EI - JP + 2SN = ES$ with a short line ___ following the mode identifier ES. If the score is positive, record it on the line provided. If it is negative, leave the mode score blank, instead recording the positive absolute value on the line for the complementary mode diagonally opposite. This complementary mode is IN for the example. A mode score of -3 for the ES mode in the upper left corner would be recorded as a positive 3 in the IN mode square below and to the right, diagonally opposite the ES square. The ES square would remain blank, the convention being that negative scores are not written down. It is understood that any blank modes represent negative scores easily obtained if needed by taking the positive score diagonally opposite and changing its sign to negative. Mode scores of zero are recorded in both the mode square and in the complementary square diagonally opposite, where the score will also be zero.

Only four such calculations are needed. It is convenient although not essential to proceed from left to right through the extraverted (E) modes in the top row. The maximum possible score for any mode is 20; scores will usually be single–digit.

There are three ways, all mathematically equivalent, to write a mode score. In the example, the positive form "IN 3" would be the most usual, although the negative form "ES -3" would sometimes be needed in an equation. A third, more awkward form would combine the variable (EI in the example) with its algebraic value, here "EI -3".

2.1.4 Mode Pattern

The scores, written in positive form separated by commas – EN 1, IN 3, ET 5, IT 2 for example – constitute the "mode pattern" for the person filling out the questionnaire. Modes with zero scores may be omitted. In this book the modes will be written in the order ES, IS, EN, IN, ET, IT, EF, IF following the order of the mode squares from upper to lower and left to right. Notice that this ordering does not necessarily put complementary modes next to each other.

The mode patterns are the key elements in team formation and organization. These can begin as soon as patterns have been collected for all members of the pool of people to be made into teams. If you are eager to do this without further discussion of the questionnaire, skip the rest of the chapter and go directly to Chap. 3 on team formation. The rest of Chap. 2 deals with psychological interpretations of the questionnaire variables in terms of Jung's personality theory, construction of the questionnaire, and proof of the questionnaire's validity.

2.2 Interpreting the Questionnaire Variables

2.2.1 Energy Direction: Jung's Attitudes Extraversion E and Introversion I

Just before World War I, the physician-psychologist Carl Gustav Jung engaged in weekly discussions in Vienna with Sigmund Freud and Alfred Adler, older founders of psychoanalysis. Watching Freud and Adler dispute whether neurosis comes primarily from outside influence or internal conflict, Jung conceived the concepts of "extraversion" and "introversion", describing Freud as an "extravert" and Adler as an "introvert". Extraversion may be regarded as the flow of psychic energy *outward* toward the exterior world. In contrast, *introversion* draws psychic energy *inward* toward one's interior psyche. These energy directions are measured by the first set of five questionnaire items. The second set of questions will partition them further into two useful subclassifications of psychic energy. Each of these subclasses

Table 2.4 Jung's cognitive modes

Extraverted Sensing $ES =$ $E + P + 2S$ **EXPERI- MENT**	Extraverted iNuition $EN =$ $E + P + 2N$ **IDEATION**	-	Extraverted Thinking $ET =$ $E + J + 2T$ **ORGANI- ZATION**	Extraverted Feeling $EF =$ $E + J + 2F$ **COMMU- NITY**
Introverted Sensing $IS =$ $I + J + 2S$ **KNOW- LEDGE**	**Introverted** iNuition $IN =$ $I + J + 2N$ **IMAGI- NATION**	-	**Introverted** Thinking $IT =$ $I + P + 2T$ **ANALYSIS**	**Introverted** Feeling $IF =$ $I + P + 2F$ **EVALU- ATION**

INFORMATION COLLECTION   DECISION MAKING

(*C*-DOMAIN)   (*D*-DOMAIN)

will be partitioned again, the first by the third set of questions; the second, by the fourth set. The eight partitions resulting will determine the cognitive modes.

Psychologists refer to introversion and extraversion collectively as "psychological attitudes", roughly following the Webster's dictionary definition of "attitude" as "... a position ... meant to show a mental state ...". Each extreme will be known as a "pole" of the "*EI* variable", terminology used throughout this book. More specifically, E and I will be called "Jung's attitudes" in this text to distinguish them from a second set of attitudes measured by the second set of questions.

Table 2.4 restates Table 2.3 to give the mode formulas in terms of single variables (e. g., E and I) rather than combined (EI) variables. In this way it shows clearly how E and I are distributed among the eight modes. The top (E) modes are said to be the "extraverted modes"; the bottom (I) modes the "introverted" ones. The adjectives "extraverted" and "introverted" are echoed in the mode designations. Equally interesting distributions of the other six variables J, P, S, N, T and F will be discussed later.

2.2.2 Information Collection and Decision-Making

Jung soon realized that his first conceptions of extraversion and introversion needed refinement. Extraversion or introversion could certainly change according to whether it was involved in collecting information or in making a decision. The psychic energy of extraversion would then be the sum of an extraverted information **C**ollection energy, symbolized by *EC*, and an extraverted **D**ecision-making energy *ED*. Introversion energy I is partitioned similarly. Note the prefix "*c-*" to abbreviate "information **C**ollection"; "*d-*" will similarly stand for "**D**ecision-making".

All variables associated with information collection, – *EC*, *IC* and others yet to be defined – will reside collectively in what will be called the "*c*-domain". For decision-making a similar "*d*-domain" containing *d*-attitudes *ED* and *ID* is defined. Notice then that whereas in Table 2.4 the extraverted *c*-attitude *ES* is at the top *left* of the *c*-domain, the *introverted c*-attitude *IN* is on the lower *right*. Similarly, *EF* is in the *d*-domain upper right whereas *IT* is lower left.

Keywords proposed for the domain attitudes are "Exploration" for *EC*, "Focus" for *IC*, "Control" for *ED* and "Appraisal" for *ID*. Be imaginative in using keywords in various contexts. For example, a person with a "control" attitude may be considered either "controlled" or "controlling" depending on the situation.

2.2.3 Orientation: Briggs Attitudes Structure J and Flexibility P

Although Jung conceived the idea of partitioning the attitudes into information collection and decision-making domain attitudes, neither he nor his followers ever devised a way to measure them as they had the attitudes I and E. Figure 2.1 suggests how this measurement might be accomplished. Notice that whereas the variable J occurs in the *extraverted decision-making* modes ET and EF, it shows up in the introverted information collection modes IS and IN. Conversely, P turns up in the extraverted *info collection* modes ES and EN and in the introverted decision-making modes IT and IF. Figure 2.1 illustrates this graphically. The J-modes are enclosed with dashed lines to contrast with the P-modes surrounded by solid lines.

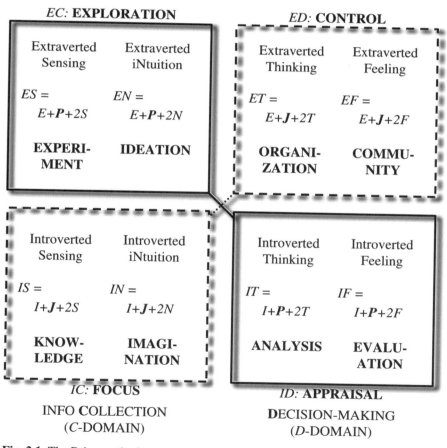

Fig. 2.1 The Briggs attitudes

It happened that Katherine Briggs, contemporary with Jung but at first not aware of his work, developed the set of questions needed. She labeled the corresponding variables "Judgment" J and "Perception" P, words that are insufficient in the current context because they do not refer to the introverted modes at all. Here they are described respectively by the keywords "Structure" and "Flexibility", which cover introverted modes as well as extraverted ones although unfortunately not mnemonic. Typologists who followed Briggs eventually realized these relations, recording verbal versions of them in Myers et al. (p. 44). In her honor these variables will be termed "Briggs" attitudes, or "orientations" to distinguish them from the earlier "Jung" attitudes E and I.

A verbal interpretation of Briggs' terminology using domain attitude keywords would regard the Judgment attitudes as "Control" (ED) for active (extraverted) decision-making and "Focus" (IC) for passive (introverted) information collection. Both involve a liking for "Structure", the keyword given jointly to these attitudes. The Perception attitudes would then be "Exploration" (EC) for active (extraverted) info collection and "Appraisal" for passive (introverted) decision-making. These both require "Flexibility", the joint keyword chosen for them here.

2.2.4 Info Collection Functions Sensing S and iNtuition N

Jung further partitioned information collection energy into two "(psychological) functions" called Sensing S and iNtuition N. These functions are complementary. Sensing uses the five senses to collect facts hands-on in the here-and-now and from the past, whereas iNtuition employs the mind to conceive present and future possibilities. Through extraversion and introversion both S and N form cognitive modes as shown in Table 2.5a. The Sensing modes ES and IS are in the left column of the info Collection domain; the iNtuition modes EN and IN make up its right column.

Notice that these new variables involve only two modes each, only half as many as the four modes touched by each of the attitudes E, I, J and P. This situation is compensated for by the coefficient 2 multiplying both S and N in the mode formulas.

2.2.5 Decision-Making Functions Thinking T and Feeling F

Finally Jung partitioned the decision-making modes according to how the decisions are made: Thinking T versus Feeling F. The Thinking way of

Table 2.5 The psychological functions

EC: **EXPLORATION**   ED: **CONTROL**

Extraverted Sensing $ES =$ $E + P + 2S$ **EXPERI-MENT**	**Extraverted** INuition $EN =$ $E + P + 2N$ **IDEATION**	**Extraverted** Thinking $ET =$ $E + J + 2T$ **ORGANI-ZATION**	**Extraverted** Feeling $EF =$ $E + J + 2F$ **COMMUNI-TY**
-	-	-	-
Introverted Sensing $IS =$ $I + J + 2S$ **KNOW-LEDGE**	**Introverted** iNuition $IN =$ $I + J + 2N$ **IMAGI-NATION**	**Introverted** Thinking $IT =$ $I + P + 2T$ **ANALY-SIS**	**Introverted** Feeling $IF =$ $I + P + 2F$ **EVALUA-TION**

IC: **FOCUS**   ID: **APPRAISAL**

a. INFORMATION COLLECTION   **b.** DECISION-MAKING
(**C** - DOMAIN)   (**D** - DOMAIN)

deciding is to consider only the non-human objects – machines, numbers, cost – among the information collected, whereas the Feeling way is to weigh human factors – people, relationships, emotions – principally. T and F can further be extraverted or introverted, giving rise to two cognitive modes for each function and accounting for the four decision-making modes. As shown in Table 2.5b the Thinking modes ET and IT are in the left column of the Decision-making domain. The Feeling modes EF and IF are in the right column. As did the information Collection functions, the Decision-making functions need a coefficient of 2 to compensate for their representation in only half as many modes as the attitudes.

2.3 Validity of the Questions

How well does the questionnaire measure the four sets of variables used to calculate the mode scores? The best way to measure them is another much longer questionnaire known as the Myers–Briggs Type Indicator or MBTI, which has been doing just this for hundreds of millions of people all over the world over the last half century. Its validity is well documented in Myers et al.

Our questionnaire is not the MBTI, although it covers the same psychological territory. It is derived from statistical studies of correlations between various MBTI questions. As described by Quenk, Hammer and Majors, these correlations establish five groupings called "facets" for each variable. From a response to any question within a facet the correlation is able to predict accurately the responses to all other items in the same facet. Each of the twenty facet correlations is in effect an abstraction of its facet questions. Each item of our questionnaire is simply a paraphrase of one of the facets discussed by Quenk, Hammer and Majors, who conveniently give names to the poles of each facet.

Take question $EI1$ for example: "You are more sociable, or (more) reserved?" This refers to the first facet of the EI variable, labeled "Initiating-Receiving" in the description on p. 24 of their book. Compare the first sentence of their Initiating description with that for Receiving.

> "Initiating: People at this pole get pleasure from mingling with others in large or small gatherings. . . . "

> "Receiving: People at this pole are much more comfortable letting conversations come to them than initiating contact. . . . "

It was then a simple matter to use "sociable" to describe "initiating" and "reserved" for "receiving" to construct item $EI1$. This approach was employed for the other nineteen facets as well. The questionnaire thus gives a facet-by-facet profile of the four variables from which the mode scores are computed.

2.4 Proof of the Mode Score Formulas

It remains to prove that the score formulas correctly transform the questionnaire responses into meaningful cognitive mode scores. Such formality may not interest students content to believe in the questionnaire's validity based on the authority of their professor; it is the professors themselves who must be convinced.

To understand the proof it helps to visualize any four questionnaire responses EI, JP, SN and TF as a point in a four-dimensional "question" space. Similarly, any four mode scores ES/IN, EN/IS, ET/IF and EF/IT can be seen as a point in a different four-dimensional "mode" space.

The formulas map response points onto mode score points. Mathematicians would call this mapping a "4 by 4 linear transformation". Since zeroes for all four responses would produce four zeroes for the mode pairs, the technical adjective "homogeneous" can be added to the description. Each set of four variables is linearly independent, making the transformation "nonsingular". This means that there is exactly one mode score outcome point for every questionnaire response input point.

All this is important because it allows us to use well-established theorems of linear algebra (Strang 1978). These say that such a 4×4 homogeneous transformation is completely determined if correspondences are given between four different pairs of response points and mode score points.

2.4.1 Four Correspondences

Aside from the origins, which for a homogeneous transformation must match automatically, there are four correspondences that come to mind. Each maximum mode score, and there are four of them, should be generated by maximum values of the related questionnaire values. For instance, the highest Extraverted Sensing ES/IN mode score should occur when the questionnaire scores for Extraversion E, Sensing S and Flexibility P are 5, their maximum values. The other three mode scores for EN/IS, ET/IF and EF/IT, needed to complete the description of the 4-dimensional mode point, should vanish for these same three questionnaire values $E = P = S = 5$, supplemented by a fourth value for the remaining variable T. Since T is not involved in the ES mode, a reasonable value for it turns out to be zero. According to the formula $ES = E + P + 2S$, the ES score corresponding to $E = P = S = 5$, $T = 0$ would be 20, making the four mode coordinates $(ES/IN, EN/IS, ET/IF, EF/IT) = (20, 0, 0, 0)$ as required.

Although the first component value $ES/IN = 20$ obviously works because it was computed from the ES formula, the vanishing of the other three must be proven to complete the match. This is accomplished by direct substitution of the questionnaire values into the three formulas for EN, ET and EF, being careful of the signs, noting particularly the new negative formula values $N = J = -5$. The calculations follow:

$$EN = E + P + 2N = 5 + 5 + 2(-5) = 0 \,;$$
$$ET = E + J + 2T = 5 - 5 + 2(0) = 0 \,;$$
$$EF = E + J + 2F = 5 - 5 + 2(0) = 0 \,.$$

Thus the formulas indeed match the mode point, where ES is maximum and the other three modes zero, to the questionnaire point where the corresponding questionnaire variables E, P and S are maximum.

The other three extreme mode points are tested similarly, the results being that $E = P = N = 5$, $T = 0$ maps to $EN = 20$ with the others 0, $E = J = T = 5$, $S = 0$ maps to $ET = 20$ with the others 0, and $E = J = F = 5$, $S = 0$ maps to $ET = 20$ with the others 0. This completes the constructive proof that the formulas map maximum questionnaire values to maximum mode scores. The 4×4 homogeneous transformation generating the mode scores is therefore completely determined, rigorously if not elegantly.

2.4.2 Other Questionnaires

Other questionnaires explore the same Jungian territory and with proper scaling can be used along with or instead of ours. The most accurate is the Myers–Briggs Type Indicator (MBTI), whose numerical "clarity" scores can be used in the same formulas as ours if the MBTI numbers are divided by 6. This scaling is needed because the MBTI numbers range from 0 to 30, whereas ours run from 0 only to 5. Our range was chosen because it generates integers (whole numbers) for the mode scores, whereas the MBTI range does not. This will of course still be true when the integral MBTI questionnaire numbers are divided by 6, but it will be seen in Chap. 3 that these decimal numbers will not cause any difficulty when forming teams.

Another usable questionnaire is the Keirsey–Bates "Temperament Sorter", which has the disadvantage that its EI range is 10 and its JP, SN and TF ranges are all 20. Since our variables all run from 0 to 5, the K–B EI score must be divided by 2 and the others by 4. The results of course may not be integers, but as for the MBTI this will not interfere with the team formation methods of Chap. 3. Keirsey also has a 100-item questionnaire on the Internet (Google: HumanMetrics), but its range is not clear at this writing. This is a problem only when other questionnaires are also in use to form teams.

2.5 Concluding Summary

This chapter has presented a Jungian cognitive mode score questionnaire to be used to make teams based on Jung's personality theory. After giving advice about a good mind-set for filling out the questionnaire, the chapter pointed out that one could use the calculated mode scores immediately to begin forming teams by the methods of the chapter following.

Intending to build the reader's confidence in the questionnaire's validity, the rest of the chapter explained and justified the questionnaire. First it briefly interpreted the four categories of questions in terms of easily measured personality descriptions. Along the way influences of the questionnaire variables on the mode scores were developed and discussed. Then it noted that the questions are based on mild abstractions of many more items from the Myers–Briggs Type Indicator (MBTI), a well-tested personality type instrument. Finally, the mode score formulas were proven by direct computation to map maximum values of the questionnaire variables to appropriate maxima of the mode scores. By the way, this proof establishes the first rigorous QUANTitative version of Jung's entirely QUALitative personality theory.

And now to make teams!

2.6 Exercises

2.1. If you are using a computerized questionnaire, compare the mode scores it gives you with those you obtain from the questionnaire in this chapter. Note any discrepancies and decide which results you prefer.

2.2. Compare your HumanMetrics questionnaire scores with those from either your computerized questionnaire or the questionnaire in this chapter. What scale factor would you apply to make them comparable? Using this factor, compute your HumanMetrics mode scores.

2.3. If you have recent MBTI results, scale them by a factor of $1/6$ and compute mode scores. Compare them to your questionnaire results, computerized or not. Which do you prefer?

Chapter 3
Team Formation by Affinity Groups

Many hands make light work.
– English Proverb, traced by Apperson to 1401

3.1 Introduction

The team formation procedure of this chapter is built upon responses by prospective team members to the cognitive mode questionnaire of the preceding chapter. Any of several questionnaires, including the well-known Myers–Briggs Type Indicator (MBTI), may be employed, but all of them must be used in the way illustrated with the Mode Questionnaire of Chap. 2. Basically, the method is to use the mode scores to partition the people into eight sets called "affinity groups", abbreviated "ag.", top scores going into the ag. s. Each ag. normally has at least as many members as there are teams, although score ties can generate a few additional members. Individuals can belong to more than one ag. Teams are formed by selecting a member from each group while honoring any non-cognitive constraints on team membership. All this can be done on a spreadsheet, as will be demonstrated. An alternative, good for collections of people who already know each other, is to have individuals choose members from the ag.s while respecting any non-psychological restrictions such as gender balance.

The chapter employs a practical example using actual cognitive patterns of twenty-one Stanford sophomores in a seminar given in 2006. The concept of affinity group is discussed and applied, eventually leading to the formation of four quartets and a quintet. The goal is to have as many modes as possible occupied by someone in the affinity group for that mode, for this guarantees that this mode score will be among the highest available. In the

Douglass J. Wilde, *Teamology: The Construction and Organization of Effective Teams* © Springer 2009

example it happens that every team but one has all eight modes covered by an affinity group member.

Attainment of such good teams is not easy; it may need computer assistance and close attention to detail. More casual approaches are also described which in the past have generated teams with at least six modes covered, already a significant improvement over random assignment or unguided personal selection by the people themselves.

The arithmetic of the situation is then generalized to any number of people and nominal team size.

3.2 Team Formation Rationale

3.2.1 Scores Measure Probability of Preference not Strength of Trait

It is tempting to regard the mode scores as in some way measuring the strength of the consciousness of the mode, or even of the ability to carry out the mode's activities well. This is not at all the case. As for the MBTI variables, the scores reflect only the probability that the subject does indeed prefer the mode rather than its complementary opposite mode (Myers et al., 1998, p, 121). Higher scores represent higher probabilities, not greater consciousness or ability.

3.2.2 High Preference Probability is Desirable

There are, however, definite advantages to a team in having each mode monitored by a person with high scores for the mode. High scores, and the resulting high probabilities, suggest that the person's interest in the mode is real and will remain significant throughout the life of the project. This is a consideration because often students will experience personality changes under the pressure of the project experience. This is all to the good for the student, who is at an age where he is finding himself, but it could upset the team's balance if he has responsibility for the mode in transition.

It makes sense then in choosing between several candidates for assignment to a mode vacancy to favor the one with the highest score. More generally, the candidates for any position should be those with the highest relevant mode scores. This idea leads naturally to the concept of "affinity group" developed next.

3.3 Affinity Grouping

3.3.1 Top Scores and Affinity Groups

It behooves the personnel pool to determine, for each of the eight modes, which people have the highest scores. Since in ideal circumstances every team should have exactly one from this top group, this group should have exactly as many members as there are teams. Since individual membership in an affinity group can depend on how many teams are involved, information on the number of teams; or equivalently the nominal team size (trio, quartet, etc.), should accompany any use of affinity groups. Identification of these top people can be done outside class on the class spreadsheet by computer or staff. Alternatively, this can be done in class by a show of hands, provided everyone is in attendance and knows their scores.

Those having the highest scores for a given mode are said to belong to the "affinity group" (ag.) for that mode. The same person can belong to more than one ag. – as many as four, or none at all. Although multiple ag. membership complicates the assignment process, it will soon be demonstrated that it also improves each team's chances to cover all their modes with high scores.

3.3.2 Marginality

Whenever the minimum score, called the "threshold (value)", for an affinity group is associated with more than one person, the ag. will actually have more members than there are teams. This happened for the *ES* ag. in the 2006 Sophomore Seminar on Teams, which had five teams. Incidentally, of the 21 students, 12 had positive *ES* scores.

Conventional ranking practice handles ties by awarding the highest of any tied ranks to all tied items, as for the third and fifth ranks in Table 3.1. Here Peg and Isabel's score of 5 earns both of them third place. Strictly speaking, they also share fourth place according to convention. Edna and Sally are similarly in fifth (and sixth) place with a score of 3. In this example membership in the affinity group requires being in the first *five* places, so all are members even though there are six of them instead of five. Since Edna and Sally share the threshold score, they are termed "marginal" members. The others are called "regular" members, including third place Peg and Isabel, whose tie does not affect their status as regulars because their common score is above the threshold value. Affinity groups with no threshold tie can consider all members regular.

Table 3.1 Five-quartet *ES* affinity group example

RANK AND SCORE

STUDENT	First	Second	Third	Fourth	Fifth	Sixth
Tom	9					
Ram		6				
Peg			5	(5)		
Isabel			5	(5)		
Edna					3	(3)
Sally					3	(3)

Table 3.2 Cognitive mode descriptions

ES **EXPERI-MENT** Discovers new ideas and phenomena by direct experience	*EN* **IDEATION** Rearranges known concepts into novel systems	*ET* **ORGANI-ZATION** Efficiently manages resources, decisive, imposes structure	*EF* **COMMU-NITY** Expressive, tactful builder of group morale
IS **KNOW-LEDGE** Physically self-aware, values practice and known technique	*IN* **IMAGI-NATION** Prophetic, guided by inner fantasies and visions	*IT* **ANALYSIS** Rationally improves quantitative performance	*IF* **EVALU-ATION** Uses personal values to distinguish good/bad
INFORMATION COLLECTION		DECISION-MAKING	

The distinction between regular and marginal membership will become important during team formation. For now, simply notice that since there are only four regular members for the five example teams, at least one team must have a marginal member totally responsible for the *ES* mode. A happy situation would have BOTH marginals on the same team instead of a regular member, for then the marginals could reasonably share responsibility according to the team role ideas of the next chapter.

3.3.3 Cognitive Mode Descriptions

Now that you've joined some affinity groups, it behooves you to know a little more about what this says about you, your potential responsibilities on your new team, and what your team-mates might expect of you. Conversely, you may wish to learn about your new team-mates, deliberately selected to be different from you. To these ends Table 3.2 gives a phrase expanding the keyword description of each of the eight cognitive modes. These phrases are abbreviations of more extensive discussions of the cognitive modes by Jungian psychiatrists June Singer (who coined the term "cognitive mode") and Mary Loomis (1984), as well as by psychologist and systems analyst H. L. Thompson (1996).

3.4 Preparation for Formation

The concept of affinity group will now be used to bring top people to every mode on every team. The cognitive patterns will be used to construct teams such that every team is composed of members who have, or are inclined to develop, interest in the cognitive modes most likely to benefit the team. At the same time, every mode must have someone's attention so that nothing is overlooked. Since the eight cognitive modes do not distribute uniformly across the pool, the team formation procedure must for fairness also randomize any particular team's risk of having modes for which no member has much interest.

3.4.1 Non-duplication

The main guideline for team formation can be summarized as "non-duplication". This means avoiding having regular members of any affinity group on the same team. The word "avoiding" is used rather than "prohibiting" because duplication could be forced whenever choices are severely limited as the last members are enlisted. Any duplication, accidental or not, would be at the expense of a clearly unfair vacancy in the mode on some other team.

The non-duplication guideline does not apply to marginal members. They are permitted on any team with a regular member of the same ag., because this is bound to happen somewhere anyway. The value of marginal members is that some of them can cover modes that would otherwise remain vacant when there are not enough regular members to service all the teams. Duplication of marginal members is in fact to be encouraged on teams having no regular members to cover the mode.

3.4.2 Team Formation Example

Material on team formation is clearer in the particular than in the abstract, so an example will be presented before generalizations are made. Consider then a personnel pool of 21 (in general p) people whose names and cognitive patterns are given in Table 3.3. The names, fictitious of course, correspond to the first 21 letters of the English alphabet, an arrangement intended to make it easier to follow what happens. The cognitive patterns are not fictitious, however; they are those of sophomore Stanford students, half heading into

Table 3.3 Cognitive patterns for the example

COGNITIVE MODES SCORES

Name	ES	IS	EN	IN	ET	IT	EF	IF
Alice	**4**	**8**					**2**	**2**
Barb		0	0	**2**	1		1	
Cora	1	3				**4**		**2**
Dana	3	**9**					**6**	
Edna	**3**	**7**			2	2		**3**
Fred*			7	**1**			1	
Gerri	1		3		1	1		**2**
Hanna		3		**1**		**6**		**0**
Isabel	**5**		**1**		0		**2**	
Jacqui	0	0	**0**	0	2		**6**	
Karen	0	4		0	**5**	**5**		
Liz	1	1			2	**6**		
Mary		4	**1**	**1**	0		**4**	0
Ned*	**4**	**7**			**7**	3		
Oprah			**1**	**1**	**3**		1	
Peg	**5**	2			**8**	2		
Qatar*	0		**8**	0		1		
Ram*	**6**	2			**4**	**4**		**3**
Sally	3	**5**					**7**	
Tom*	**9**	1			1	1		**9**
Ury*		3		**7**		1		

* INDICATES MALE

Quartet Affinity Group regular scores are marked in **boldface.**

Quartet Marginal scores are marked in **<u>underlined boldface</u>**.

engineering, in a seminar taught by the author in 2006. The problem here is to use the cognitive patterns to form teams having about four students each.

The earlier *ES* affinity group example Table 3.1 was in fact taken from the *ES* column of Table 3.3 following. The affinity groups have been formed, the results printed in **enlarged boldface** with marginal scores underscored. These numbers can of course be generated by the questionnaire computer input program Although identification of affinity group membership can also be computerized, the output is more convenient if the affinity group numbers are replaced by letter ag. labels as in Table 3.4.

In Table 3.4 following, ag. numbers have been replaced by affinity group abbreviations, and a other numbers have been removed for improved clarity. Males are marked by asterisks (∗) to help with gender balancing during team formation.

Table 3.4 5-quartet affinity groups and mode count

QUARTET AFFINITY GROUP MEMBERS   MODE COUNT

Name	ES	IS	EN	IN	ET	IT	EF	IF	REG.	MARG.
Alice	*ES*	*IS*					*EF*	*IF*	1	3
Barb				*IN*					1	
Cora						*IT*		*IF*	1	1
Dana		*IS*					*EF*	*IF*	3	
Edna		*IS*							1	
Fred*			*EN*	*IN*				*IF*	2	1
Gerri			*EN*						1	
Hanna			*EN*	*IN*		*IT*		*IF*	1	3
Isabel	*ES*		*EN*				*EF*		1	2
Jacqui							*EF*		1	1
Karen					*ET*	*IT*			2	
Liz						*IT*			1	
Mary			*EN*	*IN*			*EF*		1	2
Ned*	*ES*				*ET*				1	1
Oprah			*EN*	*IN*	*ET*				1	2
Peg	*ES*	*IS*			*ET*				3	
Qatar*			*EN*					*IF*	2	
Ram*	*ES*				*ET*	*IT*			3	
Sally		*IS*					*EF*		2	
Tom*	*ES*								1	
Ury*				*IN*				*IF*	2	

*INDICATES MALE

Affinity Group regular memberships: **boldface**.
Marginal memberships: **underlined boldface**.

3.4.3 Thresholds

Replacing the numbers of Table 3.3 with letters in Table 3.4 clarifies the team formation process to follow, but it also suppresses numerical information that could help the staff understand such global characteristics of the personnel pool as the scarcity of people favoring the iNtuition modes *EN* and *IN*. This lost information is easily recovered by noting the eight threshold values as in Table 3.5 following. Notice how high the threshold is for introverted sensing *IS*. This shows there are many scholarly Knowledge mode people in the seminar, in contrast to the few Ideation *EN* folks com-

Table 3.5 Quartet thresholds

ES	*IS*	*EN*	*IN*	*ET*	*IT*	*EF*	*IF*
4	7	1	1	3	4	2	2

Underscored thresholds are marginal.

Table 3.6 5-quartet mode multiplicity ranking

	QUARTET AFFINITY GROUP MEMBERS								MODE COUNT	
Name	*ES*	*IS*	*EN*	*IN*	*ET*	*IT*	*EF*	*IF*	*REG*	*MAR*
Dana		IS					EF	IF	3	
Peg	ES	IS			ET				3	
Ram*	ES				ET	IT			3	
Fred*			EN	IN				IF	2	1
Karen					ET	IT			2	
Qatar*			EN					IF	2	
Sally		IS					EF		2	
Ury*				IN				IF	2	
Alice		IS					EF	IF	1	3
Hanna	ES		EN	IN		IT		IF	1	3
Isabel	ES		EN				EF		1	2
Mary			EN	IN			EF		1	2
Oprah			EN	IN	ET				1	2
Cora						IT		IF	1	1
Ned*	ES				ET				1	1
Barb				IN					1	
Edna		IS							1	
Gerri			EN					IF	1	1
Jacqui							EF		1	
Liz						IT			1	
Tom*	ES								1	

* INDICATES MALE
Affinity Group qualifying scores are marked in **boldface.**
Marginal scores are marked in <u>**underlined boldface**</u>.

plementary to *IS*. Nevertheless the mentally diverse teams eventually formed performed quite well.

An alternative to disseminating Table 3.3 is to put out a threshold table like Table 3.5. Then individual students can determine their affinity group memberships directly from their own mode scores.

3.4.4 Mode Multiplicity

The two columns on the right of Table 3.4 record the number of regular and of marginal affinity group memberships. They can of course be determined by computer to save another simple but mildly tedious enumeration. This mode multiplicity information figures importantly in the strategy to be described for avoiding mode duplication while forming the teams. A final preparation step is to sort the lines in decreasing order of regular modes. Regular mode ties are resolved by further sorting in decreasing order of marginal modes. Table 3.6 is the result for the example. Further ties are resolved by the alphabetical order.

3.5 Tactics

The overall strategy of seeking to have all a team's modes assigned to affinity group members is implemented by tactics for choosing each member. The first member of a team, called the team's "seed", is drawn automatically from among the most multimodal candidates, The second member, who with the seed member forms the team's "core", is chosen to generate a joint mode pattern covering as many modes as possible, usually five or more. This necessarily draws on the multimodals remaining after the seeds have been designated, a process which, although permitting personal choice, can get complicated without computer assistance. The remaining members, being mostly unimodal, are placed easily, allowing in fact some personal choice. Everyone is eventually placed, and unavoidable variations in team size develop naturally.

3.5.1 Seed Assignment

Team identification numbers can easily be assigned to the team seeds directly from Table 3.6 because it was sorted to put the most multimodal at the top. Of the example five seeds, the first three clearly deserve their status be-

cause they are the only trimodal people on the list. The fourth person Fred*
is unambiguously chosen next as the only regular bimodal person with an
additional marginal mode bringing his total up to three as well.

The fifth seed could be any of the next four remaining bimodals Karen,
Sally, Qatar* and Ury*. The arbitrary decision rule of alphabetic precedence
is then invoked to give the Team 5 seed assignment to Karen and relegate
her fellow bimodals Sally, Qatar* and Ury* to later assignment rounds. Even
though she is one mode behind the other seeds, it will be seen that Team 5
will catch up when the second members are selected.

3.5.2 Core Dyads

Until now the automatically constructable multiplicity list (Table 3.6) has
been able to carry out the preliminaries of the team formation strategy, even
the designations of the team seeds. But at this juncture the pool faces two
ways to continue, one communal and the other computer-assisted.

The communal approach used at Stanford turns over the rest of the formation process to the students with only mild assistance from the staff. It
is left to the individuals to put themselves into teams, guided by the non-
duplication strategy and the multiplicity list.

This approach, developed for and applied to 40 freshman engineering
quintets in 2007 at Shanghai Jiao-Tong University (called "Jiao Da" by the
Chinese), focuses special attention on selecting second members to fill out
the team core dyads. At Jiao Da this took careful enumeration of many possible pairings, a task better left to a computer in the future. Despite a few mis-
takes by the eight teaching assistants who acted as "Team-meisters", each
overseeing a "Division" of about 25 students to be formed into 5 nominal
quintets, most of the teams managed to cover all eight modes and perform at
prize-winning level on simple Stanford Seminar projects. The morale boost
was remarkable, although this could have been due to the novelty of this
first experience of successful team activity on such a large scale. By the
way, translation was not needed because students at this elite university had
been studying English since the third grade.

In retrospect, there is a better, more systematic, way to look for promising cores. This is to divide up the search between the team seeds and the
unseeded multimodals, leaving the unimodals out of the process.

Each seed looks through the unseeded multimodals, noting any allowable
matches that meet or exceed a combined coverage of five modes. Regular
and marginal counts should also be noted, high regular counts being preferred. If no 5-mode matches occur, the seed must for consolation note al-
lowed matches that cover at least four modes.

Table 3.7 Core formation matrix

TEAM	1	2	3	4	5
SEED	Dana	Peg	Ram*	Fred*	Karen
Qatar*		5/0	**	**	
Sally			4/1	4/1	
Ury		5/0	**	**	
Oprah	**4/2**				
Alice			3/2		3/2
Hanna		**4/2**			
Mary		**4/2**	**4/2**		3/2
Ned*	4/1		**	**	

Only non-duplicating 5- and 6-mode matches are shown;
Numbers are regular/marginal mode counts.
** indicates two males.
6-mode entries are in **boldface**

At the same time, the unseeded multimodals conduct a similar search among the team seeds. Any matches that exceed five should of course be brought to the attention of the potential dyad partner. Although without the potential perfection of a computerized search, this procedure has a good chance of detecting every promising core pairing. The mere existence of promising pairings does not of course guarantee actual matching. One or both potential partners may have more than one promising opportunity, and non-psychological constraints such as gender balance may intervene.

If all the multimodals carry out this activity conscientiously, they could together generate the core formation matrix of Table 3.7. This matrix could be computed automatically, saving everybody a lot of effort, and at this writing such a computer code is being designed for potential academic distribution. But in more casual circumstances, the four exceptional 6-mode matches shown in **boldface** would need to be communicated directly among the seven students involved. The following casual team formation example will be described as if the matrix has not been made explicit and is merely an aid to the reader's understanding.

3.5.3 The Casual Approach

The casual approach to choosing core members would have match-ups considered simultaneously outside of class by both the seeds and the unseeded multimodals, their mutual goal being to find non-duplicated pairings covering at least five modes between the two. Since it is easy to fill the last two open positions without duplication, five-mode core dyads could count

Table 3.8 Example casual core patterns

	TEAM MEMBERS				PATTERNS					MODE COUNTS MARG. REG.	TOT.	
1	D, O	_	IS	<u>EN</u>	<u>IN</u>	ET	_	EF	IF	4	2	6
2	P, H	ES	IS	_	<u>IN</u>	ET	IT	_	<u>IF</u>	3	3	6
3	R*, M	ES	_	<u>EN</u>	<u>IN</u>	ET	IT	EF	_	4	2	6
4	F*, S	_	IS	EN	<u>IN</u>	_	_	EF	IF	4	1	5
5	K, A	<u>ES</u>	IS	<u>EN</u>	<u>IN</u>	_	IT	EF	IF	3	3	6

on covering at least seven modes by the time the team is full. As will be demonstrated, 8-mode coverage is not out of reach because of flexibility provided by the extra marginals.

In the example, Dana would find that she is unable to pair with anyone but Oprah, who in turn cannot join any seed but Dana. Fortunately this match covers six modes, four of them regularly, so the Team 1 core immediately turns out to be Dana and Oprah. Combining their mode patterns gives the Team 1 core pattern ___ IS EN IN ET ___ EF IF, repeated on the first line of Table 3.8.

Peg can cover five modes regularly by pairing with either Qatar* or Ury*, at the same time gaining a scarce male for Team 2. She might be tempted to pick one of the men immediately without looking among the women multi-modals, but Hanna and Mary inform her that pairing up with either of them would cover six modes, four regular and two marginal. It happens that Mary can make an equally good match with Ram*, at the same time satisfying the gender constraint, so Hanna remains to pair up with Peg as indicated on the second line of Table 3.8.

The three other cores complete the table. With 6 modes already covered, Teams 1, 3 and 5 can expect to cover all 8 modes when two more members are added; the other two teams should cover at least 7.

3.5.4 Later Members

With the core patterns known, it is easy for the eleven people remaining to see the teams where they might fit in. For example, ES ag. members Isabel and Tom* can see that the two openings for them are on Teams 1 and 4. Since Team 1 needs a male but Team 4 does not, Tom* would join Team 1 and Isabel Team 4, where she brings a new mode *ET*. Her marginal *EN* does not interfere with Team 4's existing *EN*. This sort of thing continues "casually" until the final teams shown in Table 3.9 are formed. This is not, of course, the only team roster possible.

As predicted, Teams 1, 3 and 5 have all eight modes covered, the consequence of having cores covering six modes. Team 4, with a 5-mode core, covers only seven modes. Barb is added to form the one quintet required. Even though she cannot cover the *IT* opening, she is assigned to the team so that her regular *IN* mode can upgrade Fred*'s marginal *IN*. By the way, being a mode short is no catastrophe. The next chapter will indeed show how this team can deal intelligently with the problem.

Although Team 2 only had a 5-mode core, it does cover all eight modes. Finally, notice that all teams have gender balance. There is a man on every team, with an unavoidable two on the quintet (Team 4).

Table 3.9 Final team roster

Team	Name				Pattern				
		ES	IS	EN	IN	ET	IT	EF	IF
1	Dana		IS					EF	IF
1	Oprah			*EN*	*IN*	ET			
1	Tom*	ES							
1	Liz						IT		
2	Peg	ES	IS			ET			
2	Hanna				*IN*		IT		*IF*
2	Qatar*			EN					IF
2	Jacqui							EF	
3	Ram*	ES				ET	IT		
3	Mary			*EN*	*IN*			EF	
3	Edna		IS						
3	Cora						IT		*IF*
4	Fred*				*IN*				IF
4	Sally		IS					EF	
4	Isabel	ES		*EN*				*EF*	
4	Ned*	*ES*				ET			
3	Barb				IN				
5	Karen					ET	IT		
5	Alice	*ES*	IS					*EF*	*IF*
5	Gerri			EN					
5	Ury*				IN				IF

* INDICATES MALE

A major advantage of the casual approach is that it allows friends, room-mates and old acquaintances to team up as long as no duplication is thereby generated. With the casual approach, everyone looks first at their friends' patterns for possible matches. Conversely, it hasn't proved difficult to avoid people with whom one is uncomfortable because there are usually several alternative ways to cover an open mode. Although the search for mental diversity may certainly bring relative strangers into contact, and it is unreasonable to expect everyone to be friendly when the team is formed, no one should be on a team with someone with whom they have had bad previous experiences or that they can't at least tolerate at the outset.

3.5.5 Team Size

A question of constant discussion among teachers of team project courses is, "what team size is best?" The preceding study of quartet team formation showed influences of team size on the number of modes covered as well as on threshold values. Table 3.10 shows that in the example, as team size increases, threshold values necessarily go up, along with the probability of stability of the affinity groups. There is less mode multiplicity, which makes it easier to construct the cores and place later members. And there are more people on the teams for covering all the modes. So mental diversity is easier to come by on larger teams. This was demonstrated convincingly on Jiao Da's casually formed forty (nominal) quintets in 2007.

On the other hand, trios may be needed when there are many different projects but only a few people. Advocates point out that trio decisions can be made more easily because there is little possibility of a tie vote. In Stanford's graduate design course, globalization has generated pairs of trios for international projects handled largely by videoconferences and e-mail. That is, each partner country provides a trio, bringing each team's total to six. It would

Table 3.10 Thresholds for various team sizes (example)

TM. SIZE	NO. TMS.	ES	IS	EN	IN	ET	IT	EF	IF
3	7	3	4	0	0	1	1	1	2
4	5	4	7	1	1	3	4	2	2
5	4	5	7	1	1	4	4	4	3
7	3	5	7	3	1	5	5	6	3

seem better to form each trio independently rather than attempt to treat the combined pair as a single sextet during team formation. Notice, however, that in the example at least, five trio thresholds would sink below 2. The two intuitive mode thresholds would in fact drop all the way to zero, where ag. membership probability is only 50%. In such circumstances, the staff must avoid pairing any two teams having zero scores for the same mode. The organization of such pairs will be discussed further in the next chapter.

3.5.6 Team Arithmetic

In any team course the staff must decide how to place a given number p of students on t teams of size either s_{min} or s_{max}, the latter being one member larger. The following inequality of integer terms expresses the relations between these quantities.

$$ts_{min} \le p \le ts_{max} \tag{3.1}$$

If s_{max} is given, t is the next integer greater than p/s_{max}. When s_{min} is given, t is the next integer smaller than p/s_{min}. Earlier in the chapter for example, $21 (= p)$ students were assigned to quartets ($s_{min} = 4$) and quintets ($s_{max} = 5$), in which case t is between $21/5 = 4.2$ and $21/4 = 5.25$, 5 teams in all.

If the number of teams t is given, the double inequality shows that s_{min} and s_{max} must bracket p/t. Had the example required 6 teams instead of 5, the two team sizes bracketing $p/t = 21/6 = 3.5$ would be trios and quartets.

Expressing p/t as an integer plus a fraction is instructive, for the integer will be the minimum team size, and if the denominator of the fraction is t, the numerator will be the number of larger teams. This numerator will be simply the "remainder" r of the long division determining the integer. Thus $p = rs_{max} + (t - r)s_{min}$. In the example for $t = 6$, $p/t = 3 + 3/6$, so there are 3 quartets and 3 trios, making $3 \times 4 + 3 \times 3 = 12 + 9 = 21 = p$, indeed the total.

3.5.7 Newcomers

In a college class or corporation, new people may arrive after teams have been formed. The least disruptive procedure for integrating the newcomers would be to assign them as extra members to existing teams of less than nominal size. There is an opportunity here for faculty or superintendent to assign newcomers so that their cognitive patterns will be of maximum value to the teams they are joining, which necessarily can reorganize slightly to take advantage of the new talent.

3.6 Concluding Summary

This chapter has demonstrated how to form teams using affinity groups based on the cognitive patterns of people in a personnel pool. Multiple affinity group membership is exploited to cover the eight cognitive modes with affinity group members, thereby guaranteeing top mode scores for every team. Each team is "seeded" with a person having high mode multiplicity. Then a second person with slightly less multiplicity is attached to each team to make the total mode coverage as high as possible. This can be done either "casually" by the people in the pool, or in a structured manner with computer assistance. The casual plan had a mode vacancy on only one team, and no duplication except for the expected number of marginal placements. Once formed, the teams are ready to be organized for task assignment in the next chapter.

3.7 Exercises

3.1. If you are a team seed, find your match(es) among the unseeded multimodals with which you can form a core dyad covering the largest total number of modes. How many modes are regular? How many marginal?

3.2. If you are an unseeded multimodal, find your match(es) among the team seeds with which you can form a core dyad covering the largest total number of modes. How many modes are regular? How many marginal?

3.3. Using the database of the example, generate a set of nominal **trios**. Discuss the results with regard to mode absence and duplication.

3.4. Using the database of the example, generate a set of nominal **quintets**. Discuss the results with regard to mode absence and duplication.

3.5. Using the database of the example, generate a set of nominal **septets**. Discuss the results with regard to mode absence and duplication.

Chapter 4
Organizing a Team

"Know Thyself" is a good saying, but not in all situations. In many it is
better to say "Know Others."
– Menander: Thrasyleon, c. 300 B. C.

4.1 Introduction

Once a team has been formed it should be organized for maximum effectiveness. This involves assigning its members to activities needed by the team
in which they are likely to excel.

Imagine being on a team whose members have just determined their cognitive patterns by the methods of the preceding chapter. The team certainly
should assemble and review the roles suggested by each person's cognitive
modes. This chapter provides graphical and tabular aids, usable in a formal team meeting, for organizing a team in this way. Presently most team
project courses leave all this to the teams themselves to handle outside of
class, formal faculty guidance ending with the formation of the teams. This
chapter argues instead for a formal organization meeting, a procedure that
has proven successful in the Stanford Sophomore Seminars and most strik-
ingly for the forty Jiao Da freshman teams in Shanghai. In hindsight, such
meetings might have raised the 73% award rate noted in Fig. 1.1, maybe
even to 100%. ALL teams of championship quality – a noble ideal that was
approached closely at Jiao Da.

Courses combining teamwork instruction with technological matter may
devote class time to the organization meeting and require submission of
a team organization plan for faculty review. Such an exercise is especially
justified in corporate situations. There the team formation process, heav-
ily constrained by limited personnel availability, special technical needs and

Douglass J. Wilde, *Teamology: The Construction and Organization
of Effective Teams* © Springer 2009

seniority considerations, is likely to produce teams that are far from ideal. When this happens it is especially important to identify and cover vacancies in the organization. This is needed to wring the highest possible effectiveness out of a team lacking full cognitive variety.

A short but serious organization meeting also overcomes a difficulty inherently brought about by the very mental variety of a well-formed team. This is the potential for dispute between pairs of members whose modes are mutually complementary and therefore represent opposing viewpoints. Management consultants in fact often look for hostile complementarity when trouble-shooting unproductive teams. Even when no open quarrel erupts, complementarity can breed unspoken distrust infecting the team.

To prevent this, faculty, supervisors and team-meisters need to warn teams of this possibility and urge team members to become aware of the advantages respect for variety brings to every complementary pair. To each team member it might be said, "Your complementary partner, by liking to do what you don't like to do, keeps the team from overlooking important matters 'under your personal radar'". Once one understands complementarity, one is less likely to take differences of opinion personally. Different people see things differently, and it is possible to disagree without being disagreeable.

Reinforcing such admonitions is a face-to-face organization meeting early in the life of the team. In such a meeting, each person becomes aware of how responsibilities probably will distribute themselves among the members. In the process, one becomes acquainted with one's complementary partners, tacitly agreeing to respect each other's differing points of view even when not agreeing with them.

Another important reason for an early organization meeting is that it speeds up the integration of the team by quickly establishing mutual communication and areas of responsibility. This need, plausible in itself, was proven experimentally by Connell and Delson in 2004. In a carefully designed double-blind experiment with 150 freshmen science and engineering students in a team project course at the University of California at San Diego, Connell assigned half the students randomly to quartets while the author assigned the other half to teams diversified cognitively by the casual methods known at the time. Neither the students nor Delson, the instructor, knew which teams were random and which cognitively diverse. Psychology professor Connell questioned the students and determined, with statistical significance, that the diversified teams took longer to coalesce than the random ones. Interestingly enough, Mechanical Engineering Professor Delson perceived the diversified teams as "more creative", also with statistical significance. Speeding up the unification of the deliberately diversified teams is therefore an important justification for an early organization meeting.

This chapter presents graphical and tabular aids to team organization and illustrates them with numerical examples adapted from Team Four of the preceding chapter. The valuable new concept of "team role", the partition of each mode into two job-like activities, is developed to handle duplications and vacancies on a team's roster.

4.2 Aids to Team Organization

4.2.1 The Team Pattern

Because so many cognitive modes are involved, it is wise to construct a chart showing from whom high preference is expected for each mode. It is also well to know if any modes might be overlooked due to lack of interest. A good instrument for this is a team mode map, constructed from a blank mode map by labeling each mode quadrant with the name of any member expected to have at least a moderate preference for it. This is illustrated for Team Four, a quintet from the Final Team Roster of Table 3.9 for the Chap. 3 example. Later examples of difficult organization problems will be generated by removing a member or two, a situation that can easily happen in university or corporation. Such truncated teams will be distinguished from this team, known as *Quintet* Four, by titles related to their team size such as *Quartet* or *Trio* Four.

Table 4.1 lists the team members, together with their score patterns from Table 3.3. The affinity groups from which the teams were constructed are no longer shown. This is because, as will be developed later, scores that are decent but not high enough to qualify for affinity group status are now of interest for filling vacancies, resolving duplications and distributing responsibilities.

Table 4.1 Cognitive patterns for quintet four

Member	ES	IS	EN	IN	ET	IT	EF	IF
Fred*			7	1			1	3
Sally	3	5					7	1
Isabel	5		1		0		2	0
Ned*	4	4			7	3		
Barb		0	0	2	1		1	

* INDICATES MALE
Scores qualifying for quartet affinity groups are shown in **boldface.**

4.2.2 Filling Vacancies

Recall that seven of Quintet Four's modes are filled by affinity group members; only Introverted Thinking *IT* is vacant. The *IT* column in Table 4.1 indeed has blanks for four of the members, but Ned* just happens to have an Introverted Thinking score of *IT*3. Although not qualifying for the *IT* affinity group, whose threshold is 4 (from Table 3.5). Ned's score of *IT*3 is certainly respectable enough to earn him the *IT* responsibility, especially considering that Barb's even smaller *IN*2 score puts her in charge of the sparsely populated *IN* Imagination mode for the team. Thus even though scores not at ag. level were not used to form teams, they still may be high enough to fill vacancies as in this example. A later Trio Four example will deal with the less fortunate situation in which a mode column is entirely empty.

4.2.3 Team Mode Map

Table 4.2 displays the assignment of these responsibilities on a mode map, for visual clarity stripped of keywords, functions and domain attitudes. It shows that Quintet Four has all eight modes covered, seven with affinity group members in the upper fifth of the pool. The eighth mode *IT* is filled adequately with Ned's score of 3, even though it is not in the top quartile.

4.2.4 Mode Duplication

Because the matching rules allow threshold and regular members to be on the same team, a mode can have more than one person in the same affinity

Table 4.2 Quintet four mode map

ES		*EN*			*ET*		*EF*	
ISABEL	5	FRED*	7		NED*	7	SALLY	7
NED*	4	Isabel	1	-	Barb	1	Isabel	2
Sally	3	Barb	0		Isabel	0	F* &, B	1
IS		*IN*			*IT*		*IF*	
SALLY	5	BARB	2		Ned*	3	FRED*	3
Ned*	4	Fred*	1	-			Sally	1
Barb	0						Isabel	0

* INDICATES MALE
Affinity Group members are shown in
upper case boldface.

group. Quintet Four's Extraverted Sensing ES mode has both regular ag. member Isabel ($ES5$) and threshold member Ned* ($ES4$). In principle then, Isabel and Ned* could share the responsibilities for Experimentation.

A more formal solution would split the mode into two jobs or "roles", in this case the "Tester" who, to quote Table 3.2, "pushes the performance envelope hands-on" and the "Prototyper" who "builds models and prototypes". The two individuals, in a discussion with the other team members, could decide who should do what. The personality-based role concept (Wilde 1999) will be developed more fully later in the chapter.

Alternatively the team could decide to give both roles to the same person. Since Table 4.2 shows that ES is Isabel's only ag. mode, whereas Ned* has two others as well (ES and the non-ag. mode IT), it would seem reasonable to give her both roles to allow Ned* more time for his other responsibilities.

To make things even more complicated, Sally's $ET3$ could in principle also put her in line for the ES responsibility. But like Ned* she has two other modal responsibilities (IS and EF), so Isabel would still seem the most likely ES candidate.

4.2.5 Organization Map

The assignments resulting from the team's organization meeting are conveniently summarized on the cleaned up mode map of Table 4.3. It shows only the final assignments, scores and other names being removed to improve clarity. Notice that affinity group membership is not displayed, it no longer being relevant once the teams have been formed.

Table 4.3 Quintet four organization

ES	EN		ET	EF
ISABEL	FRED	-	NED*	SALLY
IS	IN		IT	IF
SALLY	BARB	-	NED*	FRED

* INDICATES MALE

4.2.6 More Difficult Organization

Because of the careful way in which Quintet Four was formed, as were the four other teams in the Chap. 3 example, its organization was quite easy. This is often not the situation when the formation process is less disciplined or when members leave the team. Professionals may be transferred, and students are usually free to drop even a project course. Organization can also be complicated by late newcomers to the team. To generate a more complicated organization problem, consider what would happen if both Ned and Sally were to leave the team, leaving Trio Four with the mode map shown in Table 4.4. Sally and Ned took four affinity group memberships with them (*IS*, *EF*, *ET* and *IT*), and now neither Knowledge *IS* nor Analysis *IT* definitely appeals to anyone. How to repair these new vacancies in the organization is the topic of the sections immediately following.

4.2.7 Reaching for a Vacant Mode

The *IS* Knowledge mode vacancy is relatively easy to fill. Barb's *IS* 0 score indicates indifference rather than active dislike, so she is the obvious candidate. Moreover, *IS*'s adjacency to Barb's *IN* assignment indicates that she already has the focused introverted *IC* information collection attitude. Thus shifting the focus from possibilities to facts once in awhile should not be too hard for her. The team's Knowledge responsibility, usually involving searching the literature and the Internet, will join her *IN* imaginative sketching and strategic duties.

Table 4.4 Trio four mode map

ES	*EN*		*ET*	*EF*
ISABEL 5	**FRED** 7	-		
	Isabel 1		Barb 1	Isabel 2
	Barb 0		Isabel 0	Barb 1
				Fred 1
IS	*IN*		*IT*	*IF*
	BARB 2	-		**FRED** 3
	Fred 1			Isabel 0
Barb 0				

Affinity Group members are shown in
upper case boldface.

Filling the *IT* Analysis mode vacancy is a little trickier because everyone in the trio professes a preference for the complementary Community mode *EF*. But this preference is indeed slight, Barb and Fred scoring only *EF* 1, numerically equivalent to *IT* − 1. The solution then is to give them each an *IT* role, Barb's adjacent to her new mode assignment *ET* Organization and Fred's next to his *IF* Evaluation assignment. Such assumption of a slightly disfavored role or mode will be known as "reaching". This sensible idea was originally suggested by the psychiatrist Dr. John Beebe, at the time President of the San Francisco Jungian Institute. The reached-for roles, which will be described later in more detail, are depicted in the *IT* mode square of Table 4.5.

4.2.8 Mode Sharing

Table 4.5 also shows partition of modes *IN* and *ET* because scores for the two members preferring each mode differed by only one insignificant point (*IN*: B2 vs. F1; *ET*: B1 vs. I 0). As discussed earlier, each person's role was placed next to that person's mode, fortuitously possible everywhere in this example.

By coincidence this plan also distributed the roles as equally as possible; six to Fred and five to both Barb and Isabel. Shrinking the team from five to three only added one role to Fred's already large responsibility, but the two women each took on three new roles. By the way, the roles added, having slight or even slightly negative scores, were not responsibilities the members would have expected before joining the team. It was the team's needs rather than the members' interests that generated these particular as-

Table 4.5 Trio four role map

ES	*EN*		*ET*	*EF*
ISABEL	FRED	-	ISABEL	ISABEL
			BARB	
IS	*IN*		*IT*	*IF*
	FRED		BARB	
BARB		-		FRED
	BARB		FRED	

signments. It may be comforting to know that, despite such adversities as the loss of two members, the team can generate a reasonable organization plan with no holes in it.

4.3 Team Roles

The concept of team role was first suggested by Belbin (1993). The present personality theory context appeared in Wilde 1999, which also used erroneous transformation mathematics corrected here in Chap. 2. This slight error does not affect the role theory.

The two Extraverted Sensing roles Tester and Prototyper were defined in the discussion of the quintet's mode duplication (Sect 4.2.4). The fourteen remaining will be now be developed.

4.3.1 Jungian Team Role Keywords

Table 4.6 shows a full role map with the modes separated from each other to allow simultaneous display of keywords for the functions and domain attitudes. This bisection of the modes yields sixteen roles, each given a "keyword" in the form of a job title for some activity on an engineering design team. Although some of the keywords may be less appropriate for legal, medical, journalistic or military teams, professionals should have little difficulty translating them as needed.

Table 4.6 Role map with keywords

TESTER		ENTRE-PRENEUR		COORD-INATOR		DIPLO-MAT
Es	-	En	-	Et	-	Ef
eS		eN		eT		eF
PROTO-TYPER		INNO-VATOR		METHOD-OLOGIST		CON-CILIATOR
-	-	-	-	-	-	-
INVESTI-GATOR		VISION-ARY		SPECIAL-IST		NEED FINDER
iS	-	iN	-	iT	-	If
Is		In		It		If
INSPEC-TOR		STRAT-EGIST		RE-VIEWER		CRITI-QUER

INFORMATION COLLECTION   DECISION-MAKING

Consider the fourteen role keywords not already discussed in connection with the quintet. In the information collection domain the EN Ideation mode yields the more EC exploratory "Entrepreneur" En role and the more N possibility-oriented "Innovator" eN role. Notice that both roles are given unique identifying letter pairs En and eN. The capital in each pair indicates whether the attitude $((E+P)/2$ for mode $EN)$ or the function (N for EN) is larger. The IS Knowledge mode has the more IC focussed "Inspector" Is error-detecting role as well as the more S fact-oriented "Investigator" iS literature-searcher role. The IN Imagination mode includes the more IC focussed "Strategist" In future-oriented role as well as the more N possibility-oriented "Visionary" iN visualization role.

Among decision-making modes the ET Organization mode partitions into the more ED controlling "Coordinator" Et for line organization and the more T objective "Methodologist" eT for technical staff matters. The EF Community mode has two roles: the ED controller "Diplomat" Ef and the F people-oriented "Conciliator" eF who detects and mediates interpersonal differences. The IT Analysis mode partitions into the more ID appraising "Reviewer" It and the more T technical "Specialist" iT. Finally, the IF Evaluation mode has two roles: the ID evaluating "Critiquer" If and the more people-oriented iF "Need-finder".

Table 4.7 gives short descriptions of the sixteen team roles. Knowing these roles helps each member understand what s/he can do well for the team and what to expect from the team-mates.

4.3.2 Advantages of the Role Formulation

Although subdivision of modes into roles increases the amount of terminology, it has two advantages to justify it, one organizational and the other semantic. The organizational advantage of role subdivision is the power and flexibility it gives to the team planning process. This has already been demonstrated in the discussion of Trio Four.

The semantic advantage is that the roles are personifications of responsibilities, easily viewed as approximate job descriptions. This brings the somewhat abstract cognitive modes down to earth by expressing them in terms of personal activities responsible to the team. For example, it is usually easier to view oneself as an Innovator and Entrepreneur than as a leader in using the Ideation cognitive mode EN. Studying the range of team roles helps a person understand the complex needs of a problem-solving team and of his/her own expected contribution.

Table 4.7 Team role descriptions

Es **TESTER** Pushes performance envelope hands-on	*En* **ENTRE- PRENEUR** Explores and promotes new products and methods	*Et* **COORD- INATOR** Focuses activities to save time and effort	*Ef* **DIPLOMAT** Harmonizes team, client and consumer
eS **PROTO- TYPER** Builds models and prototypes	*eN* **INNO- VATOR** Synthesizes new products by component modification	*eT* **METH- ODOLOGIST** Sets deadlines, modifies procedures and breaks bottlenecks	*eF* **CONCIL- IATOR** Detects and resolves interpersonal issues
iS **INVESTI- GATOR** Gets facts and know-how about prior experience	*iN* **VISIONARY** Visualizes unusual designs forms and uses	*iT* **SPECI- ALIST** Analyzes performance and efficiecy	*iF* **NEED FINDER** Evaluates human factors and consumer issues
Is **INSPECTOR** Detects errors and enforces specifications	*In* **STRAT- EGIST** Speculates on project and product future	*It* **RE- VIEWER** Compares performance to goals and standards	*If* **CRITI- QUER** Addresses aesthetic and moral issues

INFORMATION COLLECTION   DECISION-MAKING

4.3.3 Efficacity

As motivation for the reader to examine the role concept further, some pertinent quotations from Stanford Psychology Professor A. Bandura's 1997 book *Self-Efficacy: The Exercise of Control* (New York, Freeman) follow. On p. 3 Bandura defines, in the context of sociocognitive psychology, "perceived self-efficacy" as "beliefs in one's capabilities to organize and execute the courses of action required to produce given attainments." Eric Adamson, after taking a Stanford Sophomore seminar on teams followed a year later by Professor Bandura's course, suggested that understanding one's own team roles should certainly enhance one's perceived self-efficacy in these roles. This is important for every team member because, as Bandura continues, "Such beliefs influence the course of action people choose to pursue, how much effort they put forth in such endeavors, how long they will persevere in the face of obstacles and failures, their resilience to adversity, whether their thought patterns are self-hindering or self-aiding, how much stress and depression they experience in coping with taxing environmental demands, and the level of accomplishments they realize."

On p. 38 Bandura continues, "a resilient sense of efficacy enhances sociocognitive functioning in the relevant domains in many ways. People who have strong beliefs in their capabilities approach difficult tasks as challenges to be mastered rather than as threats to be avoided. Such an affirmative orientation fosters interest and engrossing involvement in activities. They set themselves challenging goals and maintain strong commitment to them. They invest a high level of effort in the face of failures and setbacks. They remain task-focused and think strategically in the face of difficulties. They attribute failure to insufficient effort, which supports a success orientation. They quickly recover their sense of efficacy after failures or setbacks. They approach potential stressors or threats with the confidence that they can exercise some control over them. Such an efficacious outlook enhances performance accomplishments, reduces stress, and lowers vulnerability to depression. These findings offer substantial support for the idea that beliefs of personal efficacity are active contributors to, rather than inert predictors of, human attainments." How's that for motivation to master one's team roles?

An organization meeting does more than increase the efficacy of the individual team members. It also raises the team's *perceived COLLECTIVE efficacy*, which Bandura defines (p. 477) as "a group's shared belief in its conjoint capacities to organize and execute the courses of action required to produce given levels of attainment." Such potential improvements in personal and group efficacy are powerful arguments for early team organization.

4.3.4 The Team Pattern Map

Figure 4.1 is a pattern map for Trio Four shaded to display how each member covers combinations of roles which might be called "meta-roles" – abstrac-

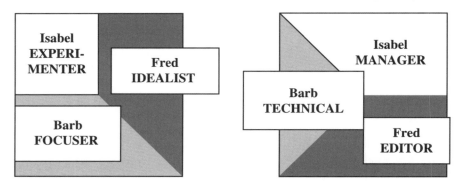

Fig. 4.1 Pattern map for trio four

tions that simplify description of each member's responsibilities and thus clarify the overall "team pattern" distinguishing it from other teams.

Fred's six roles are split equally between intuitive "Idealist" (*En* Entrepreneur, *eN* Innovator and *iN* Visionary) and appraising "Editor" (*iF* Needfinder, *If* Critiquer and *iT* Reviewer). Barb's two *T*-roles in objective decision-making (*eT* Methodologist and *iT* Specialist) earn her the adjective "Technical" to modify her three focused *c*-roles *iS* Investigator, *Is* Inspector and *In* Strategist. Isabel's three control roles *Et* Organizer, *Ef* Diplomat and *eF* Conciliator earn her the responsibility of Manager. The *c*-mode remaining – Isabel's *ES* Experimenter – can keep its modal keyword.

The trio then is composed of Isabel the hands-on manager, Fred the idealistic editor and Barb the technical focuser. Between them they cover all of Jung's cognitive modes. These new descriptors give useful labels for the individual members that can help them understand what each other is supposed to do. Such existential job titles are only suggestive examples here; in general they are best left to the imagination of the team members themselves.

The team's organization meeting can end now that every role has been covered and the teams's personality made clear not only to the members, but to supervisor, faculty and/or coaches.

The Trio Four example went more smoothly than do many team organizations, perhaps because the trio was a subset of Quintet Four, which was put together by the careful methods of Chap. 3. More casually constructed teams may have ragged patterns with several possibilities for improvement that each need to be discussed by the assembled team before final assignments are made. By the way, it is not essential to distribute the roles equally or have a person's roles adjacent to each other. These properties of the Trio Four organization pattern may have been coincidental.

4.3.5 Other Team Sizes

The number of members on a team does not have to be four as in the example. Team size can vary with project needs, but some general ideas can be stated about how size affects a team's pattern. The larger the team, the higher will be maximum clarity scores for a given mode, which in itself should improve team performance. The educational downside is that the average number of roles per person will necessarily decrease, meaning that student members will have less chance to develop a range of roles.

Quintets come under consideration for very large classes supervised by a small faculty and staff, simply because there are fewer teams to watch over. This was the size of the 40 Shanghai freshman teams who did so well in 2007.

With class cognitive distributions like those at many American universities, where there is a relatively large proportion of *ET* Organization modes, it is tempting on large teams to introduce a hierarchical organization with one "foreman", selected either by the team or the faculty. Being the main intermediary between the team and its client or supervisor, the foreman would usually favor the duplicated *ET* organization mode and could easily oversee the team's organization plan. Another *ET* would then take the usual responsibilities on the team for coordination and/or methodology.

Professional diversity or international communications considerations may call for even larger teams. Currently each Stanford graduate design team has three Stanford members working in remote contact with three colleagues at an overseas university.

4.4 Concluding Summary

This chapter has discussed the actual organization of a team into informal task areas, an activity that can be carried out either formally in a supervised meeting or informally as the project proceeds. Two planning aids were developed, the team role map and the team pattern map. The team role map is a quick graphic scheme for allocating responsibilities, resolving role duplications, and covering low consciousness roles. The pattern map displays the team's unique identity for the edification of supervisor, faculty or coach, as well as for the team itself. Meeting to generate these documents is an important starting point for any team project work. Bandura's concept of efficacity, personal as well as collective, is shown to be relevant to the team role concept.

4.5 Exercises

Exercise 4.1. Carry out the detailed mode and role analysis of Quartet Four, obtained by having Fred leave Quintet Four.

Exercise 4.2. Do the detailed mode and role analysis of Quartet One from the team roster of Table 3.9.

Exercise 4.3. Carry out the detailed mode and role analysis of Trio One, obtained when Oprah leaves Quartet One of Table 3.9.

Chapter 5
Personal Description

This above all – to thine own self be true:
And it must follow, as the night the day,
Thou canst not then be false to any man.
– Polonius, in Shakespeare's *Hamlet* (1601)

5.1 Introduction

Past chapters have used cognitive mode patterns to form and organize teams, showing when individual team members need at times to employ portions of their mode patterns, sometimes having to reach into relatively undeveloped modes, for the good of the team. This chapter helps an individual understand his/her own particular cognitive pattern, independently of the needs of any particular team. It does this by showing how to assemble descriptions composed by Myers for the famous sixteen-element MBTI Personality Type Table. But whereas Myers only used one description per person, the novel approach here is to allow multimodal people to combine two such descriptions.

This chapter goes beyond what is needed to form effective teams – all that many project courses require. The material following is intended rather for courses, often at first- or second-year college level, which go beyond teamology to focus on development of the team-members as individuals. Since 2002 earlier versions of this material have been used in the Stanford Sophomore Seminar "Creative Teams and Individual Development", and its immediate predecessor was developed for the entire freshman engineering class of the Shanghai Jiao-Tong University (Jiao Da) in 2007. Students of college age have a natural interest in who they are and what they might become. This chapter helps them understand what their questionnaire responses can tell them.

The chapter opens with a review of Myers' theory of "personality types". This type theory is different from that of the Myers–Briggs Type Indicator (MBTI) as an instrument intended to measure a subject's preferences for each pole of the four paired categories E vs. I, etc. Chapter 2 showed how to transform the MBTI numbers into cognitive mode scores useful for forming and organizing teams. The present chapter deals with how Myers associated a personality type, along with its description, with each of the sixteen combinations of the four pairs of letters, $ESTJ$ for example. Since she didn't use the associated numbers, Myers had to make certain assumptions, some but not all originated by Jung. This led her to associate each type precisely with a particular pair of cognitive modes. Knowing these assumptions, one can reverse the process, first determining the modes by the cognitive formulas of Chap. 2 and then deducing the types afterwards from the modes.

Myers type descriptions involve exactly two cognitive modes, whereas a cognitive pattern can have as many as four significant scores or even none at all. Thus the cognitive mode approach augments Myers' theory by allowing a secondary Myers description to supplement the usual one, to be described here as the "primary type", determined by the letters alone. This multiple-type approach is demonstrated on several Stanford sophomore example patterns from Table 3.3. Of the twenty-one students, nine come up with the single primary type of the Myers theory. Two more have double primary types arising from function ambiguity as with the Myers approach. The other ten have configurations involving a secondary Myers type, itself occasionally also doubled due to function ambiguity.

This combination of the cognitive mode approach with Myers type descriptions can predict and guide an individual's healthy psychological development. This is demonstrated suggestively by a student example based on retesting at the end of the seminar after several projects have been designed and built. The example also shows how team-based learning can develop an individual participant. Further anecdotes recount how the multiple-type approach has guided past students to better type descriptions and away from false ones.

5.2 Myers Personality Type Descriptions

The MBTI, mentioned in Chap. 2 as an alternative and indeed best possible input to the mode formulas of the Cognitive Questionnaire, primarily estimates a subject's preference for each pole of the four pairs E vs, I, S vs. N, T vs. F and J vs. P, in order slightly different from that of the cognitive questionnaire of Chap. 2. In practice the four letter preferences, $ESTJ$ for example, direct the subject's attention to a short verbal type description like the one following (Myers and McCaulley 1989, p. 20). All questionnaires except that in Chap. 2 give the four letters explicitly. When needed, the letters can be recovered from the cognitive mode scores as will be shown in Sect. 5.5.

5.2.1 ESTJ

Practical, realistic, matter-of-fact. Decisive, quickly move to implement decisions. Organize projects and people to get things done, focus on getting results in the most efficient way possible. *Take care of routine detail.* Have a clear set of logical standards, systematically follow them and want others to also. Forceful in implementing their plans.

This description, along with fifteen associated with the other letter combinations, was composed by Isabel Myers when she and her mother Katherine Briggs published the MBTI. Although influenced of course by Jung's descriptions of the three categories E vs. I, S vs. N and T vs. F, as well as those of Briggs concerning J vs, P, Myers in fact constructed each of the $2 \times 2 \times 2 \times 2 = 16$ type descriptions by pairing cognitive modes in particular ways. The pairs were selected according to three guidelines.

The first guideline, due to Jung, was that one mode must be taken from the four information collection modes; the other, from the four decision-making modes. The second, due to Myers, was that the two modes must have different attitudes, one introverted and one extraverted. Finally, Jung stated that one mode, called the "dominant" mode, is more influential than the other "auxiliary" one lending support.

Table 5.1 shows how Myers associated the 16 mode pairs resulting with the 16 letter types. For example, the $ESTJ$ type (number 13) is associated with dominant mode ET, listed first, and auxiliary mode IS in the second place. The parts of the $ESTJ$ description coming from the auxiliary mode IS are shown in *italics* in the preceding description. The types have also been numbered here to aid reference to the Myers type descriptions on pp. 20-1 of (Myers and McCaulley 1989). Myers did not need a numbering system because her scheme was entirely based on the letter combinations. In contrast, the system to be developed here will be based primarily on mode pairs, leading to Table 5.2 in the next section to be explained as it is put to use.

5.3 Primary Type Descriptions

From now on the Myers types will be referred to collectively as "primary" to contrast them with other "secondary" types deduced later from the complete modal pattern. The section begins by discussing a precise procedure for focusing attention on the larger "significant" mode scores while neglecting the slight ones. Then a student example will be presented in which the mode pattern leads exactly to a single Myers type. A second student example will show how uncertainty in one of the function preferences leads to a double-type involving two Myers types, both of which must be considered mutually primary.

5.3.1 Score Significance

This section is based on p. 122 of Myers et al. (1998). MBTI clarity scores range from 0 to 30, scores below 6 being regarded as "slight" – liable to swing either way upon later testing. Scores from 6 through 15 are called "moderate"; from 16 through 25 "clear"; higher scores "very clear". The modal theory developed here will distinguish only between the slight scores and all the others, which will be collectively considered "significant". Since the questionnaire formulas of Chap. 2 range up to 20 instead of 30, the lower bound for significant mode scores will be taken as $(20/30) \times 6 = 4$. Hence mode scores of 3 or less will be considered "slight".

Although a slight mode cannot, from this point of view, be considered dominant, it may fill the auxiliary position when no other mode follows the Myers guidelines. Slight modes, although not confirming preference, do in-

Table 5.1 Type table with cognitive mode patterns

ATTITUDE PAIR	ST	SF	NF	NT
IJ (ID)	1. *ISTJ* *IS , ET*	2. *ISFJ* *IS, EF*	3. *INFJ* *IN, EF*	4. *INTJ* *IN, ET*
IP (ID)	5. *ISTP* *IT, ES*	6. *ISFP* *IF, ES*	7. *INFP* *IF, EN*	8. *INTP* *IT, EN*
EP (EC)	9. *ESTP* *ES, IT*	10.*ESFP* *ES, IF*	11. *ENFP* *EN, IF*	12. *ENTP* *EN, IT*
EJ (ED)	13. *ESTJ* *ET, IS*	14. *ESFJ* *EF, IS*	15. *ENFJ* *EF, IN*	16. *ENTJ* *ET, IN*

dicate inclinations that may be useful for estimating secondary or potential types.

5.3.2 Single-Type Example

As an example of how to determine a primary type, consider Fred from Quintet Four of Chap. 3. His modal pattern from Table 3.5 is *EN* 7, *IF* 3, *IN* 1, *EF* 1. The mode with the largest score is *EN*, and since this is significant take *EN* as the dominant mode. Of the three modes remaining, *IN* cannot be auxiliary because it is an information collection mode like *EN* (Jung's first guideline). Moreover, *EF* has the same extraverted attitude as *EN*, in violation of the second guideline due to Myers. Hence only *IF* can qualify as Fred's auxiliary mode.

Table 5.2 has all this reasoning built into it and so can be used to find Fred's primary type corresponding to *EN*, *IF*. In Table 5.2 read across the "*EN* dominant" line to find an entry with column heading matching some other of Fred's modes, in this case *IF*. The table entry *ENFP* is seen to be the corresponding "primary type". Since the remaining two mode scores are both less than 4, there is no need to look for a secondary type.

On Quintet Four Sally (*EF* 7, *IS* 5, *ES* 3, *IF* 1) is also a single primary-type (*ESFJ*), as is Isabel (*ES* 5, *EF* 2, *EN* 1, *ET* 0, *IF* 0) (*ESFP*). In the 21-student sophomore seminar, six more (G, L, M, O, Q and T) are single primary-type – nine in all.

5.3.3 Double Primary-Type Example

There are also situations in which two Myers types may with equal plausibility both be primary – the double primary-type case. For example, on

Table 5.2 Converting mode pairs into letter type

	ET	*IT*	*EF*	*IF*
ES dom.		9. *ESTP*		10. *ESFP*
ES aux.		5. *ISTP*		6. *ISFP*
IS dom.	1. *ISTJ*		2. *ISFJ*	
IS aux.	13. *ESTJ*		14. *ESFJ*	
EN dom.		12. *ENTP*		11. *ENFP*
EN aux.		8. *INTP*		7. *INFP*
IN dom.	4. *INFJ*		3. *INFJ*	
IN aux.	16. *ENFJ*		15. *ENFJ*	

Quintet Four Barb (*IN* 2, *ET* 1, *EF* 1, *IS* 0, *EN* 0) has no significant mode scores. But rather than being without a type, she can use her *IN* 2 to indicate a potentially dominant "leading" *IN* mode. In Table 5.1 the *IN* 1 line has a tie in both columns *ET* and *EF*. Hence table entries *INTJ* and *INFJ* are both plausible primary types for Barb because of her ambiguity in preferring a decision-making function. This double primary-type situation can be written either *INXJ* or *INTJ–INFJ*. Along with one other double primary-type (Jacqui), this brings the total number of students fitting the Myers model to exactly eleven out of twenty-one

Barb's case is interesting because her slight scores generated a double type covering two Myers descriptions. In this way ambiguous preferences can indicate type multiplicity perhaps resolvable later as the personality develops.

5.4 Secondary Type Descriptions

5.4.1 Double-Type Example

Of the remaining ten students in the seminar, seven (A, D, E, K, N, P and R) have a secondary type backing up the primary type. Either or both types can in principle be doubled. As an example of a person with one primary and one secondary type, consider Quintet Four's Ned (*ET* 7, *ES* 4, *IS* 4, *IT* 3), whose dominant mode is clearly *ET*. Table 5.2 must be employed differently this time because the dominant is a decision-making mode involving the columns. In column *ET* the only auxiliary row matching some other of Ned's modes is *IS* auxiliary, which makes the table entry *ESTJ* the primary type.

Ned's remaining modes *ES* 4, *IT* 3, which can only correspond to the dominant and auxiliary modes respectively, generate *ESTP* as the secondary type here. Strictly speaking, it is exaggerated to use the words "dominant" and "auxiliary" for the secondary type, but it is not worth the trouble to introduce more terminology here as long as the words are understood to be merely aids to using Table 5.2.

Ned then has a secondary type *ESTP* backing up his primary type *ESTJ*. Table 5.1 shows that although these types differ only by one letter, they together involve all four modes of Ned's cognitive pattern. This is an important consideration when assembling teams. Thus Ed could if needed cover the *ES* Experiment and/or the *IT* Analysis modes even though these modes are not part of his *ESTJ* primary type description. Indeed, Table 4.3 shows that Quintet Four did assign Ned to the *IT* associated with his secondary type *ESTP* rather than the *IS* associated with his primary type *ESTJ*. All for the good of the team.

5.5 Type Letters from Mode Scores

5.5.1 Exceptions

So far eighteen of the twenty-one seminar students have been accounted for. The other three (C, H and U) appear at first not to fit either the one- or the two-type models described. It happens though that reasonable fits can be obtained by working with the letters rather than the modes. Ury's case is the most difficult because all four modes in his pattern happen to be introverted (as are Hanna's), apparently leaving no openings for the Myers guideline that the attitudes of the dominant and auxiliary be different. Myers' letter approach will, however, generate a plausible extraverted mode to balance Ury's description. This will be demonstrated after it is shown how to find the letters corresponding to a given modal pattern, Ury's for example.

5.5.2 Letters from Modes

Figure 5.1 shows the eight cognitive modes, each mode square containing the appropriate mode formula from Chap. 2. The dotted arrows indicate how

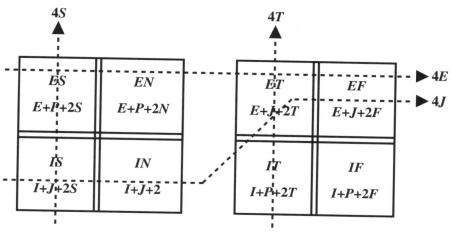

Fig. 5.1 Recovering clarities from mode scores

certain formulas can be added together to recover the original questionnaire counts from which the modes were calculated. For instance, adding together the top four extraverted modes gives

$$ES + EN + ET + EF =$$
$$(E + P + 2S) + (E + P + 2N) + (E + J + 2T) + (E + J + 2F) = 4E .$$

The radical simplification occurs because the complementary variables (P, J), (S, N) and (T, F) cancel each other out in the sum, as suggested at the point of the arrow. Recall that any or all of these extraverted modes can be negative. Such negative mode scores are obtained from the corresponding complementary positive mode score used for team construction simply by changing the sign. A glance at the illustrative numbers in Fig. 5.2 may clarify this.

Similar cancellations generate the expressions for $4S$, $4T$ and $4J$ shown at the other three arrowheads. A negative value for any of them simply indicates that the complementary variable has the same value with a positive sign.

The factor 4 is just an artifact of the mathematical conventions employed in Chap. 2 to guarantee that only integers would be generated. Dividing any result by 4 gives the whole number cognitive questionnaire score E, etc., generating the corresponding sum of the mode scores. Recall that these questionnaire numbers range only up to 5, whereas MBTI clarity numbers are bounded above by 30, a consideration important only if the two sets of scores are to be compared. For the present purpose, however, the actual numbers are not important. Only the signs, which determine the letters, are needed for the analysis to follow.

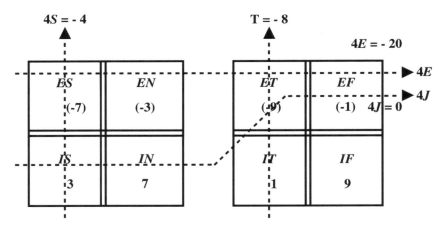

Fig. 5.2 Example: Ury *IF* 9, *IN* 7, *IS* 3, *IT* 1

5.5.3 Letter-Type Example

For a numerical example instructive in itself because of its usefulness in a case not amenable to the preceding procedures, consider Ury with his four introverted modes IF 9, IN 7, IS 3 and IT 1. As diagrammed in Fig. 5.2, summing the extraverted modes gives $4E = -12$, $4S = -8$, $4T = -4$ and $4J = 0$, which determines the Myers letter type as the ambiguous $INFX$. Thus Ury has a double primary-type $INFJ–INFP$.

The corresponding modes from Table 5.1 are IN and EF for $INFJ$ together with IF and EN for $INFP$. Since IF has its mode score slightly higher than that for IN, it would perhaps be more accurate to write the double type with $INFP$ first, i. e., $INFP–INFJ$. Thus the extraverted modes EN and EF can be considered small "reaches" from Ury's original modes IS and IT. This idea was confirmed in a retest described in the next section on Personal Development.

Incidentally, Cora's IT 4, IS 3, IF 2, ES 1 generates the primary type $ISTP$ with IT dominant and ES auxiliary. Hanna'S (IT 6, IS 3, IF 2, IN 1) letter type $ISTP$ has IT dominant with auxiliary function S, with Myers' attitude guideline requiring it to be extraverted, a very slight reach from IN 1. Along with Ury, Cora and Hanna bring the total number of students satisfying the Myers model to thirteen out of twenty-one. Of the eight double-types remaining, seven have correct matches between their primary-types and the MBTI letters. Karen the exception is discussed next.

5.5.4 A Misleading Letter-Type

Trimodal Karen (ET 5, IT 5, IS 4, ES 0, IN 0) has $ISTJ$ for her letter type, its dominant mode IS and auxiliary mode ET counter to the actual score order ET 5, IS 4. This anomaly is verified by Karen's primary-type $ESTJ$ and secondary-type $ISTP$ obtained by modal analysis. Since the type description for $ISTJ$ is devoted mainly to IS rather than the larger-scored ET, it is a bit misleading, especially since IT is not mentioned at all.

The point here is that, although mostly reliable, the MBTI letter type can mislead occasionally. Never accept it without making a confirmatory modal analysis.

5.6 Personal Development

5.6.1 Ury's Evolution

After eight weeks of designing and building half-hour hands-on projects in the Stanford seminar, Ury retook the questionnaire, going from the old pattern (*IF* 5, *IN* 4, *IS* 2, *IT* 1) to a new one (*IF* 9, *IN* 5, *EF* 3, *IS* 1). The most noticeable change was his reach from *IT* 1 to *EF* 3, his first extraverted mode to emerge. A similar change was the near doubling of the dominant *IF* score from 5 to 9.

His letter type both before and after was the ambiguous *INFX*, expressible as the double type *INFP–INFJ*. Participation in the team projects seems to have increased scores for all four of Ury's modes associated with the double-type, the changes being D*IF* = 4, D*IN* = 1, D*EF* = 4, D*EN* = 1. These were generated by a decrease of two questionnaire points for *I* and an increase of one point for *F* as Ury became less introverted and even more people-oriented than when he began the seminar.

5.6.2 Attitude Balance

Although Ury's change is in the right direction, his top two modes *IF* and *IN* are still both introverted, in violation of Myers' attitude balance guideline. This calls for reflection because this guideline, although an imperfect predictor of the auxiliary, does express conditions that Jungians like Dr. John Beebe of the San Francisco Jungian Institute consider conducive to good mental health in the long run. Faced with a major discrepancy between empirical and rule-based personality description, one can only benefit from understanding not only how the difference may have arisen but also whether and how it might be reduced. The next two sections discuss experience with this guideline in the seminar.

Dr. Beebe told the 2003 seminar that before he experienced Jungian therapy to cure a reading block after medical school, his leading modes were introverted: *IN* and *IT*. In analysis he realized that his information collection mode should be *EN*, as it is now. His explanation was that as a child he had introverted his intuition to please his *IN* mother. Part of the inspiration for extraverting his *IN* mode to *EN* was Isabel Myers' guideline as expressed in *Gifts Differing* (Myers, I. and Myers, P. 1980) Beebe credits this principle as a brilliant ideal of type development, novel, bold and badly needed when introduced.

When Beebe spoke, the two-type model developed in this chapter had not yet been conceived. In hindsight one can speculate what the two types would have been for him. Three of Beebe's modes are accounted for by his lecture – *IN*, *IT* and eventually *EN*. If as a dutiful son Beebe emulated his mother's dominant *IN*, it is plausible that he also adopted *ET* to please his father, a forceful senior Army officer. Putting his parental imitations together, with *IN* dominant, would yield the Myers type *INTJ*, whereas the modes *EN*, *IT* of his later type give *ENTP*. A reasonable description of his personality transition would be from a youthful *INTJ* primary and *ENTP* secondary to the mature *ENTP* primary and *INTJ* secondary, a simple type interchange.

Attending that seminar was Eric Adamson, who two years later introduced Bandura's concept of *self-efficacity* to the course as described in Chap. 4. A double-extravert himself (ET^*, *ES*) at that moment, Eric was so impressed with Beebe's story that he began to question the origin of his own doubly-extraverted auxiliary *ES*. He realized that his *ES* came from sub-conscious emulation of his ES^* father, a famous surgeon. He soon saw that a more natural auxiliary mode for him was *IN*, the complement of *ES*. His later questionnaire results indicated *ET*, *IN*, the attitudes balanced in accord with the Myers guideline. Thus his originally false primary-type indicator *ESTJ* is now *ENTJ*.

Unlike Eric, a majority of the few seminar students violating the Myers guideline are quite comfortable with the fact. Moreover, they have been effective as assigned to their cognitively diversified teams according to the leading modes they reported on the questionnaire. Who knows how many of them will change their auxiliary modes as time goes by and they reflect on past influences?

5.6.3 Words from a Double-Introvert

Readers steeped in MBTI tradition may still doubt that double-extraverts and double-introverts really exist and are healthily productive. To strengthen the case for double-version and its value when placed properly on a cognitively diverse team, here is a lightly edited excerpt from the 2003 term paper of Soph Seminar member Brandon Bunke-Quintero, all of whose positive modes, like Ury's, were introverted. His eight-mode quartet, rare before the development of the theory in Chap. 3, was without a doubt the best until then in the seminar, and he was one of its multiple stars.

"I never developed a penchant for working in teams in high school and junior high. I preferred to work by myself in a quiet setting where I could focus. I have never been very loud nor do I like to attract attention. Most of the teams I worked with were deeply flawed. Not taking into account certain personality types while making teams can have dire consequences and reduce productivity dramatically."

"One typical problem was when two leaders would fight over which ideas would take precedence. The two egomaniacs would make so much noise because they wanted to be heard that the rest of the group members would be muffled out. The few strong extraverts hindered my performance; I got shoved aside."

"Unfortunately, I started to think maybe I was different from others, or that other people were simply rude and not considerate of my ideas. After learning about team dynamics and variation in personality types, I understand why I had so much trouble on teams. ... A malformed team can lead to undue stress, lack of productivity, and the isolation of individuals with thoughts of doubt and social ineptitude. ..."

"In terms of team performance, I was fortunate in this class to be placed on a team that dominated almost every event we participated in. During construction, nobody got into arguments, and whenever somebody had a different idea it was analyzed and used if everybody agreed on it. ... Working in such a conducive environment increases the probability of future success because the teammates are more likely to be willing to work again using the method."

Just to complete the picture, Brandon also had a doubly extraverted teammate. The two of them together sparked the team, whose other two members satisfied the Myers guideline.

5.7 Concluding Summary

This chapter has shown how to use Myers' sixteen type descriptions to portray a person's cognitive pattern of four cognitive mode scores. The new approach is to use, when appropriate, a secondary type description in addition to Myers' single primary one based on only two of the modes and defined entirely by the four questionnaire letters but not by any clarity scores. Slightly over half the seminar sample fit the Myers' single type model, the others requiring a secondary description as well. The new broader approach shows promise for correcting false types and guiding a person's future type development. An example showed how participating in hands-on projects on a cognitively varied team can generate noticeable productive personality changes.

5.8 Exercises

5.1. For your mode pattern, what are your primary and secondary types?

5.2. Find the primary and secondary types for Alice (*IS* 8, *ES* 4, *EF* 2, *IF* 2). What is her letter type?

5.3. Find the primary and secondary types for Jacqui (*IF* 6, *IT* 2, *ES* 0, *IS* 0, *EN* 0, *IN* 0). What is her letter type?

5.4. Find the primary and secondary types for Tom (*ES* 9, *IS* 1, *ET* 1, *IT* 1). What is his letter type?

Chapter 6
Innovations and Errors

A theory should be simple – but not too simple.
– Albert Einstein

This final chapter summarizes, on three levels, what has been said. First it highlights eight innovations that carry individual typology into the new realm of teamology – the technology of constructing and organizing effective teams. These are not contributions to psychology. They follow from straightforward systems analysis of teams.

Second it narrates the development of teamology over the past two decades, including the errors and accidental discoveries that have advanced the subject. The purpose of this existential log is to build confidence in the impact of the preceding innovations on team effectiveness, even though not all the underlying principles have yet been scientifically proven. Third it draws attention to unsolved problems and unproven techniques in order to motivate ongoing teamology research in engineering, business and psychology.

6.1 System Analysis Innovations

Although the innovations to follow involve elements of psychology, they mainly concern the logic and simple mathematics of systems analysis. The underlying psychological principles are untouched; the novelties being in how these principles are combined on teams. The main advance is bringing quantitative reasoning to a personality theory that has until now largely avoided using its own numbers.

Douglass J. Wilde, *Teamology: The Construction and Organization of Effective Teams* © Springer 2009

6.1.1 Transformation of MBTI Clarities
into Jungian Cognitive Mode Scores

By far the most important contribution of this book is the set of eight transformation equations in Tables 2.3 and 2.4. The transformation bridges the
gap between individual personality assessment and the formation and organization of teams. It also has two mathematical advantages; it is arithmeti-
cally simple, requiring only algebraic addition and multiplication by two.
And it produces scores that are integers – whole numbers – with no fractions or decimals to slow things down. Perhaps of equal value, theoretically
at least, is that its validity is proven rigorously by direct computation as
shown in Sect. 2.4.

6.1.2 Affinity Grouping

Another innovation for guiding and simplifying team formation is the affinity group concept. Using relative ranking rather than absolute mode scores
adapts successfully to whatever mode distribution the personnel pool experiences. At the same time it de-emphasizes unimportant score differences. It
allows every team to have the top people in the pool assigned to every one
of the eight cognitive modes.

6.1.3 Team Formation Tactics

Recognizing that individuals can belong to several affinity groups at once
leads naturally to starting each team with a "seed" covering as many modes
as possible. Adding a second member who raises the total number of modes
covered as high as possible usually puts this "core" in a position to add
single-mode members later who can bring the total number of modes as
close to eight as possible.

6.1.4 Pattern Matching

The matching matrix introduced in Sect. 3.5.2 is a useful new idea for combining two cognitive patterns, whether of individuals or of collections of
several people. Although this can be tedious without computer assistance

when the pool is large, pattern matching greatly increases the chances of 8-mode coverage for all teams.

6.1.5 Filling All Vacancies

A novel way to cover an otherwise vacant mode is simply to use score priority, that is, to select the member with the highest numerical mode score, whether or not it qualifies for affinity group membership. When this highest score is negative, such an assignment is called "reaching", a good idea from psychiatrist John Beebe (Sect. 4.2.7).

6.1.6 Team Roles

The new concept of team roles (Sect. 4.3) has two applications. One is to partition tasks between two members in the same affinity group. The other is simply to amplify each mode description by subdividing it into two "roles", each associated with a rather specific job to be done on behalf of all.

6.1.7 Secondary Myers Types

The brand new notion of a "secondary" Myers type supplementing the usual single type, now called "primary", is an important extension of traditional individual typology (Sects. 5.3 and 5.4). This development reconciles potential philosophical differences that could arise between the cognitive mode approach needed for teamology and the single type description of the MBTI.

6.1.8 Abstraction of Personality Variables

A minor novelty is the Cognitive Questionnaire's use of abstract Step II definitions instead of tested MBTI items (Sect. 2.3). Although not exploited here, this approach has the potential to isolate what might be called "micro-personality" influences, as when an E person has several I responses. This concept could well extend or complement the idea of secondary Myers type.

6.2 Errors and Discoveries

The theory presented here evolved over a period of almost two decades. In addition to the eight rigorous innovations just discussed, other ideas that emerged from direct experience can only be described anecdotally. The history log following is intended to reinforce belief, at least tentative, in the overall positive impact of this book's approach upon the effectiveness of student academic teams.

6.2.1 Creativity Workshop

In 1989, colleague James Adams introduced the MBTI to a two-week Creativity Workshop held at Stanford's Design Division and supervised by Bernard Roth, Rolf Faste and the author. His point was that people are different and their differences should be taken into account when speaking of creativity, especially on teams.

Intrigued by this notion, the author went to the second edition of the MBTI Manual (Myers and McCauley 1989, pp. 214-5) and learned of a quantitative "Gough Creativity Index" (GCI) constructed empirically as a linear function of the four MBTI variables (Gough 1981). The MBTI was then administered to the 1990 and 1991 workshops to discover that, with statistical significance, the workshop experience did improve the average GCI of the twenty-five or so participants, mostly professors of engineering design (Wilde 1993).

6.2.2 MBTI Prize-Winners

Convinced then of the validity of the MBTI numbers, the author asked colleague Larry Leifer to have his graduate mechanical engineering design students be guided by the MBTI when forming their year-long project teams. No advice was offered them other than that they try to have different letters whenever possible.

The results were spectacular! The Stanford teams, only duets because of a shortage that year of students relative to projects, won ten Lincoln Foundation Design awards out of the twelve offered nationally. Over the dozen years preceding, Stanford teams had never received more than four Lincoln awards, so Leifer then made the questionnaire a standard required feature of the course. And the author began the research described in this book as an informal retirement project.

6.2.3 Creativity Index for Trios and Quartets

Figure 1.1 graphs the continual improvement of the Stanford teams in the last decade of the millennium. For the five years following the 1991 break-through just recounted, the teams were formed, not by the MBTI letters, but by distributing the highest GCI scores among the teams. Starting with 1991, over the first six years this technique doubled the fraction of teams, now trios and quartets, receiving Lincoln awards. In hindsight this procedure lent variety only to the information collection modes, entirely ignoring decision-making.

6.2.4 Unintentional Relapse

During the author's absence from Stanford in 1995, team formation reverted to the very informal procedures of pre-questionnaire days in which people picked team-mates on proximity rather than personality. The results were depressingly predictable. The fraction of teams winning prizes dropped from one-half back to one-quarter while the number of team problems rose annoyingly. This was uncomfortable proof of the new ideas for making teams.

6.2.5 Full Modal Variety

Consequently Professor Mark Cutkosky, who had just taken over the course from Leifer, reinstated the MBTI-style procedures, now using an abbreviated 20-item questionnaire that was computerized for easy student use by teaching assistant Mike McNelly. The McNelly code also incorporated the newfound principle of using all eight cognitive modes, thus seeking complete cognitive variety. That is, not only were the information Collection c-modes distributed, the Decision-making d-modes were also varied. The prize frequency immediately rose to three-quarters, a tripling of the base frequency of the preceding bad year, which matched that of the years before the new method. Although not scientific proof of the new cognitive variety approach, it was strong anecdotal evidence encouraging further exploration.

6.2.6 Error and Website Closure

After the three years of this comfortably high prize frequency, the method used to calculate the mode scores was found to be theoretically in error, although not by much numerically. Just for the record, the mistake was to treat the numbers as components of a vector in Euclidean space, an easy blunder for an engineer to make. Thus for example, this incorrect approach would have the extraverted c-attitude computed by the Pythagorean root mean square

$$EC^2 = (E - I)^2 + (J - P)^2$$

instead of the correct simple average

$$[(E - I) + (J - P)]/2 \, .$$

Mathematically speaking, the Jungian four-dimensional personality space is *non-Euclidean*!

Although this error of using root mean squares instead of simple averages still generated a large improvement in prize frequency, the website was closed. The new website now under construction will compute the modes correctly by the formulas of Tables 2.3 or 2.4. In hindsight it is plausible that the prize-winning fraction would have been higher than three-quarters were it not for this error. Unfortunately, this mistake still pops up from time to time from bootlegged websites, so BEWARE!

6.2.7 Publication of the Team Role Concept

The team role concept of Sect. 4.3 (Wilde 1999), was incorporated into the McNelly website. Unfortunately the mode computation error also appeared in that article. Two subsequent attempts to correct the error in the same journal were rejected because the whole subject had become too complicated to be treated in a short journal article. So again, beware! This book is presently the only place in the literature to find the correct transformation.

6.2.8 Course Split in 2000

In the year 2000 the course split into two as the old Design Division spun off a new Biomechanics Division with its own graduate project course. Lincoln prize frequencies no longer were useful measures of team effectiveness, which is why Fig. 1.1 does not extend into the third millennium. The introduction of overseas partnerships also limited the usefulness of the Lincoln awards as effectiveness indicators.

6.2.9 Freshman and Sophomore Seminars

Since 2003 the author has conducted at Stanford an annual three-unit quarterly seminar entitled: "Creative Teams and Individual Development" for four or five quartets of Freshmen and/or Sophomores. Having full control of the course, the "team-meister" was able to use the seminar as a laboratory for the latest teamology techniques. In addition, the seminar's individual development issues lead to the secondary Myers type concept of Sects. 5.3 and 5.4.

6.2.10 Connell–Delson Statistical Study

In 2004 Professors Joan Connell and Nathan Delson, respectively of the Psychology and Mechanical Engineering Departments of the University of California at San Diego, ran a double-blind test of the team formation techniques of the time. As subjects they used their 168 freshmen in a robotics-based hands-on project course lasting a semester. Half the teams were diversified cognitively; the other half were assigned at random, and neither Delson the lecturer nor the students themselves knew which teams were cognitively diverse. As recounted in Sect. 4.1, Connell found, with statistical significance, that although the diversified teams took longer to cohere, they ultimately appeared more "creative" to Delson. These findings motivated the team organization meeting recommended in Chap. 4.

6.2.11 Team Organization Meeting

This team organization meeting concept was adopted in 2004 by Professor Yong Se Kim of Korea's Sungkyunkwan University for his multinational "A3" teams blending students from Korea, China and Japan. The Stanford Seminar described in the preceding section first mandated organization meetings in 2005 just before the third project. For the first time that quarter, and indeed for the first time since the seminars began, all four teams produced prize-quality projects. The following year the organization meetings were held even before the first project, with the happy result that all teams generated prize-quality projects three times in a row.

Then in 2007 all the techniques described in this book were tried out on the entire 195-student freshman engineering class at "Jiao Da", the Shanghai Jiao-Tong University, allied with the University of Michigan. Remarkably, their forty nominal quintets almost universally generated projects of the same prize-quality as the Stanford Sophomore Seminar teams. This was a massive validation of the new teamology.

6.2.12 Other Universities

Readers inclined towards networking may be interested to know of colleagues at other universities with some experience in teamology at various levels, professions and numbers of students. In chronological order more or less they are:

- Carnegie-Mellon: Arthur Westerberg (Chemical Engineering emeritus), Jon Cagan (Mechanical Engineering and Business), John Wesner (All engineering specialties plus technical writing).
- Oregon State: Christine Ping Ge (120 M. E. undergraduates)
- U. of California, Berkeley: Alice Agogino (M. E. Design and Business)
- U. of Texas, Austin, Kris Wood (Mechanical Engineering)
- U. of Florida: Keith Stanfill (Management Science and multidisciplinary advanced projects)
- Loyola U. of Los Angeles: Dorota Shortell (more than 100 undergraduate engineers)
- Sungkyunkwan U., Su Won, Korea: Yong Se Kim (600 Korean freshman engineering and science majors, tri-national Asian teams)
- Shanghai Jiao-Tong and University of Michigan: Shen-Sheng (Samson) Zhang (195 Chinese freshmen engineers).

6.3 Research Needs

The preceding innovations and empirical history of teamology still leave many questions unanswered. Plausible anecdotal ideas need to be tested scientifically where possible, both by the engineers and business people who use teams, and by the psychologists and sociologists who strengthen the underlying theories.

6.3.1 Statistical Significance in Large Classes

Connell and Delson led the way in comparing teamological methods with random or other ways of assigning students to teams. Similar approaches could in principle be applied to studies of the more recent methods outlined in this book. There are two key problems: good objective measures of effectiveness and comparison with control groups.

6.3.2 Team Formation Algorithms

The team formation scheme of Sect. 3.5 is a procedure, not an algorithm with proven convergence and optimality properties. This calls for the rigorous theories of Operations Research.

6.3.3 Micro-Personalities

A secondary Myers type has letter differences from the associated primary type. When can these be explained by the Step II distinctions incorporated into the Cognitive Questionnaire of Sect. 2.3?

6.3.4 Efficacity

Bandura's (1997) individual and team efficacity ideas of Sect. 4.3.3 are readily observed. Formal psychological study would seem in order.

6.3.5 Reaching

Reaching for a complementary mode is sometimes necessary, as pointed out in Sect. 4.2.7. Is this beneficial, harmful or neutral to the reacher? How effective is it from the team's standpoint?

6.3.6 Student Development

Working on cognitively varied teams seems especially beneficial to the students involved. This would seem a fertile area for formal educational and psychological research.

6.3.7 Archetype Theory

Jungian psychiatrist John Beebe has advanced a theory assigning dramatic Jungian archetypes – witch, trickster and daemon for example – to those modes not used in the team formation process (Harris 1996). Would these ideas be usable to predict or analyze interpersonal problems on teams? Or, for that matter, individual personal problems affecting team performance adversely?

6.3.8 Midlife Transition

Can personality insights gained by students in team courses guide and smooth their later midlife transitions? To what degree do personality discoveries in a college project course predict changes in later life at midlife and beyond?

6.3.9 Descriptions

The mode and role descriptions of Tables 3.2 and 4.7 could no doubt be improved by professional psychologists. This would remove the current dependence on the two-mode Myers type descriptions that can unintentionally confuse behavioral research. It would seem to pay off if past research involving types were redone with modes alone.

6.3.10 Mode Distribution

It would be interesting to know how multimodal personalities are distributed. What fraction of various populations are unimodal? Bi- and tri-modal?

Double-verted? Does multimodality change with age? Are some patterns more stable or healthier than others?

6.3.11 Direct Mode Determination

It would seem possible to develop a forced-choice questionnaire measuring the cognitive modes directly instead of estimating them from MBTI clarities. An attempt in this direction was made by Singer and Loomis (1984), who ranked mode preferences among all eight possibilities. Simple choices between complementary modes would be easier and more direct, perhaps revealing non-linear relationships between MBTI variables and mode scores.

6.4 Concluding Summary

The fledgling discipline Teamology lies at the fuzzy interface between engineering, management science, systems analysis and psychology. Its innovations – transformation, construction technique and personality theory – have been based largely on systems analysis. Its short history is a catalog of trial-and-error opportunities exploited, not always correctly or even efficiently. There remain many interesting interdisciplinary research questions, so intriguing to both designers and psychologists that some of them like Connell and Delson might even agree to collaborate on them.

Take it from here, ye team-meisters!

Appendix

Table A.1 Individual mode and team role maps

$$ES \quad IS \quad EN \quad IN \quad ET \quad IT \quad EF \quad IF$$

Name:

Score:

INDIVIDUAL COGNITIVE MODES

ES	EN	ET	EF
Experiment	Ideation	Organization	Community
Knowledge	Imagination	Analysis	Evaluation
IS	IN	IT	IF

TEAM NAME:

MODE SCORES

Member names: *ES IS EN IN ET IT EF IF*

TEAM ROLES

Tester	Entrepreneur	Coordinator	Diplomat
Prototyper	Innovator	Method-ologist	Conciliator
Investigator	Visionary	Specialist	Needfinder
Inspector	Strategist	Reviewer	Critiquer

Glossary

Keywords are Capitalized and symbols *italicized*

affinity group (ag.)	a collection of people sharing the top scores of some cognitive mode. There are as many members as there are teams to be formed *t*
Analysis (*IT*)	Introverted Thinking *d*-mode; "Appraisal" of or by "Objects"
Appraisal (*ID*)	Introverted **D**ecision-making *d*-attitude
assignment, mode	see mode assignment
assignment, role	see role assignment
attitude	a position meant to show a mental state
attitude, Briggs	(*J/P*) Judgment/Perception
attitude, domain	*EC*, *IC*, *ED* and *ID*
attitude, Jung (*I/E*)	Introversion/Extraversion
bimodal	see modal, bi-
c-domain	information **C**ollection domain: *EC*, *IC*, *S* and *N*
clarity	preference clarity index (pci)
clarity, moderate	any preference clarity index from 6 to 15
clarity, slight	any preference clarity index of 5 or less
clear	any preference clarity index from 16 to 25
clear, very	any preference clarity index of 26 or more
cognitive pattern	see pattern, cognitive
Community (*EF*)	Extraverted Feeling *d*-mode; "Control" of or by "People"
Conciliator (*eF*)	more feeling *d*-role of *EF* Community mode
Control (*ED*)	Extraverted Decision-making *d*-attitude

Coordinator (*Et*)	more extraverted *d*-role of *ET* Organization mode
core	the first two members of a team being formed
Critiquer (*If*)	more introverted *d*-role of *IF* Evaluation mode
d-domain	decision-making domain: *ED*, *ID*, *T* and *F*
Diplomat (*Ef*)	more extraverted *d*-role of *EF* Community mode
domain	information **C**ollection (*c*-) or **D**ecision-making (*d*-)
double extraversion	having both leading modes extraverted
double introversion	having both leading modes introverted
duplication	having two members of the same affinity group on a team
dyad	two associated members who do not constitute a team
efficacity, perceived self-	beliefs in one's capabilities to organize and execute the courses of action required to produce given attainments (Bandura)
efficacity, perceived collective	a group's shared beliefs in its conjoint capacities to organize and execute the courses of action required to produce given levels of attainment (Bandura)
Entrepreneur (*En*)	more extraverted *c*-role of EN Ideation mode
Evaluation (*IF*)	Introverted Feeling *d*-mode
Experiment (*ES*)	Extraverted Sensing *c*-mode: "Exploration" of "Facts"
Exploration (*EC*)	Extraverted (info) **C**ollection attitude
Extraversion (*E*)	an attitude in which a person directs his interest to phenomena outside himself (*E*)
extraversion, double	see double extraversion
Exterior	keyword associated with extraversion (*E*)
Feeling (*F*)	*d*-function for making decisions by or about "People" (keyword)
Focus (*IC*)	Introverted (info) Collection (*IC*) *c*-attitude
Group, Affinity	see Affinity Group
Ideation (*EN*)	Extraverted iNtuition *c*-mode; "Exploration" of "Possibilities"

Imagination (*IN*)	Introverted iNtuition *c*-mode; "Focus" on "Possibilities"
Innovator (*eN*)	more iNtuitive *c*-role of *EN* Ideation mode
Inspector (*Is*)	more introverted *c*-role of *IS* Knowledge mode
Introversion (*I*)	an attitude in which a person directs his interest to his own experiences and feelings (*I*)
introversion, double	see double introversion
Interior	keyword associated with introversion (*I*)
iNtuition (*N*)	*c*-function for collecting "Possibilities" (keyword)
Investigator (*iS*)	more sensing *c*-role of *IS* Knowledge mode
Judgment (*J*)	a Briggs attitude encompassing domain attitudes *ED* and *IC*
Knowledge (*IS*)	Introverted Sensing *c*-mode; "Focus" on "Facts"
map, team	a diagram of eight squares, one for each cognitive mode, each of which may be divided into two team roles
marginal membership	see membership, marginal
membership, marginal	possessing a modal score at the threshold level of an affinity group
membership, regular	possessing a modal score above the threshold level of an affinity group
Methodologist (*eT*)	more objective *d*-role of *ET* Organization mode
modal, bi-	a cognitive pattern with exactly two significant modes
modal, multi-	a cognitive pattern with three or more significant modes
modal, non-	a cognitive pattern with no significant modes
modal, quad-	a cognitive pattern with exactly four significant modes
modal, tri-	a cognitive pattern with exactly three significant modes
modal, uni-	a cognitive pattern with exactly one significant mode

mode assignment	a mode assigned to a particular person on a team, indicated by a circle or square around the clarity value
moderate (clarity)	see clarity, moderate
multimodal	see modal, multi-
Needfinder (*iF*)	more people-oriented *d*-role of *IF* Evaluation mode
non-duplication	having no more than one member of the same affinity group on a team
Organization (*ET*)	Extraverted Decision-making *d*-mode; "Control" of or by "Objects"
pattern, cognitive	the signed clarity values of a person's four cognitive modes, the usual order being *ES*, *EN*, *ET*, *EF*
pattern, team	a table of cognitive patterns, one line for each team member
Perception (*P*)	a Briggs attitude encompassing domain attitudes *EC* and *ID*
pole	an extreme of a complementary pair of attitudes or functions
pool, personnel	a collection of people from which teams are to be formed. The number in the pool is symbolized by *p*
preference clarity index (pci)	the score for an MBTI variable or the computed equivalent for a cognitive mode or domain attitude
Prototyper (*eS*)	more practical *c*-role of *ES* Experiment mode
quadmodal	see modal, quad
reach	to assume a role or mode complementary to one's actual mode
regular membership	see membership, regular
Reviewer (*iT*)	more objective *d*-role of *IT* Analysis mode
role, team	a partition of a mode into two parts, each representing a team task. There are sixteen roles in all, each listed in the Glossary
role assignment	a role assigned to a particular person on a team, indicated by placing the person's name or initials in the proper triangular region of the team map
Sensing (*S*)	*c*-function for collecting "Facts" (keyword)

seed	the first member of a team being formed
significant (clarity)	having a clarity of 6 or greater
slight (clarity)	see clarity, slight
Specialist (*iT*)	more objective *d*-role of *IT* Analysis mode
Strategist (*In*)	more introverted *c*-role of *IN* Imagination mode
team map	see map, team
team pattern	see pattern, team
team role	see role, team
team size	the number of people on a given team
team size, nominal	when a personnel pool is being formed into teams, the approximate team size sought; symbol *s*. Actual team sizes will differ from this nominal unless the pool size is an exact multiple of *s*
Tester (*Es*)	more extraverted *c*-role of *ES* Experiment mode
Thinking (*T*)	*d*-function for making decisions by or about non-human "Objects"
threshold	minimum modal score for an affinity group
triad	three associated members who do not constitute a team
trimodal	see modal, tri-
type, double-	a type having one letter unclear
type, letter	same as type, Myers
type, Myers	one of the sixteen personality types described by Myers
type, primary	the clearer type of a type pair
type, secondary	the less clear type of a type pair
type, single-	a primary type with no associated secondary type
type description	see description, type
type pair	two Myers types associated with the same person
unimodal	see modal, uni-
very clear	see clear, very
vacancy	a role or mode to which no one on a team has been assigned
Visionary (*iN*)	more iNtuitive *c*-role of *IN* Imagination mode

References

Bandura A (1997) Self-Efficacy: The Exercise of Control, New York, Freeman.

Belbin R M (1993) Team Roles at Work, Oxford, England, Butterworth Heinemann.

Gough H G (1981) Studies of the Myers–Briggs type Indicator in a Personality Assessment Institute, paper presented at the Fourth National Conference on the Myers–Briggs Type Indicator, Stanford University, CA.

Harris A S (1996) Living with Paradox: An Introduction to Jungian Psychology, San Francisco, CA., Brooks/Cole.

Jung C G (1921, 1971) Psychological Types, Princeton, NJ., Princeton University Press.

Keirsey D, Bates M (1978) Please Understand Me: Character & Temperament Types, Del Mar, CA, Prometheus Nemesis Books.

Martin C R (1992) Looking at Type: The Fundamentals, Gainesville, FL, Center for Applications of Psychological Type, Inc.

Myers I B, McCaulley M H (1989) Manual: A Guide to the Development and Use of the Myers–Briggs Type Indicator, Palo Alto, CA., Consulting Psychologists Press.

Myers I B, McCaulley M H, Quenk N L, Hammer A L (1998) MBTI Manual: A Guide to the Development and Use of the Myers-Briggs Type Indicator, 3rd Edition, Palo Alto, CA., Consulting Psychologists Press.

Myers I B, Myers P B (1980,1995) Gifts Differing: Understanding Personality Type, Palo Alto, CA., Davies-Black.

Quenk N L, Hammer A L, Majors M (2001) MBTI Step II Manual, Mountain View, CA, CCP, Inc.

Singer J, Loomis M (1984) Interpretive Guide for the Singer-Loomis Inventory of Personality, Palo Alto, CA, Consulting Psychologists Press.

Thompson H L (1996) Jung's Function-Attitudes Explained, Watkinsville, GA., Wormhole Publishing.

Wilde D J (1993) Changes Among ASEE Creativity Workshop Participants, J. Engineering Education, **82**. 3.

Wilde D J (1997) Using Student Preferences to Guide Design Team Composition, ASME Transactions (Design Engineering Technical Conference Proceedings, Albuquerque, NM.), paper DETC/DTM-3980, New York, American Society of Mechanical Engineers.

Wilde D J (1999) Design Team Roles, ASME Transactions (Design Engineering Technical Conference Proceedings, Las Vegas, NV.), paper DETC/DTM-99 003, New York, American Society of Mechanical Engineers.

Index

Printing: Krips bv, Meppel, The Netherlands
Binding: Stürtz, Würzburg, Germany